Margarita Spirova

**Discrete Geometry in Normed Spaces**

AF061492

Margarita Spirova

# Discrete Geometry in Normed Spaces

Südwestdeutscher Verlag für
Hochschulschriften

**Imprint**
Any brand names and product names mentioned in this book are subject to trademark, brand or patent protection and are trademarks or registered trademarks of their respective holders. The use of brand names, product names, common names, trade names, product descriptions etc. even without a particular marking in this work is in no way to be construed to mean that such names may be regarded as unrestricted in respect of trademark and brand protection legislation and could thus be used by anyone.

Publisher:
Südwestdeutscher Verlag für Hochschulschriften
is a trademark of
Dodo Books Indian Ocean Ltd., member of the OmniScriptum S.R.L Publishing group
str. A.Russo 15, of. 61, Chisinau-2068, Republic of Moldova Europe
Printed at: see last page
**ISBN: 978-3-8381-2617-3**

Zugl. / Approved by: Chemnitz, TU, Hab., 2010

Copyright © Margarita Spirova
Copyright © 2011 Dodo Books Indian Ocean Ltd., member of the OmniScriptum S.R.L Publishing group

# Preface

This work refers to ball-intersections bodies as well as covering, packing, and kissing problems related to balls and spheres in normed spaces. A quick introduction to these topics and an overview of our results is given in Section 1.1 of Chapter 1. The needed background knowledge is collected in Section 1.2, also in Chapter 1. In Chapter 2 we define ball-intersection bodies and investigate special classes of them: ball-hulls, ball-intersections, equilateral ball-polyhedra, complete bodies and bodies of constant width. Thus, relations between the ball-hull and the ball-intersection of a set are given. We extend a minimal property of a special class of equilateral ball-polyhedra, known as Theorem of Chakerian, to all normed planes. In order to investigate bodies of constant width, we develop a concept of affine orthogonality, which is new even for the Euclidean subcase. In Chapter 2 we solve kissing, covering, and packing problems. For a given family of circles and lines we find at least one, but for some families even all circles kissing all the members of this family. For that reason we prove that a strictly convex, smooth normed plane is a topological Möbius plane. We give an exact geometric description of the maximal radius of all homothets of the unit disc that can be covered by 3 or 4 translates of it. Also we investigate configurations related to such coverings, namely a regular 4-covering and a Miquelian configuration of circles. We find the concealment number for a packing of translates of the unit ball.

Based on this work and on the recognition of the corresponding teaching qualification, the author obtained the academic degree DOCTOR RERUM NATURALIUM HABILITATUS (Dr. rer. nat. habil.) from the Chemnitz University of Technology on the 2th of December 2010.

# Contents

**1 Introduction**   1
    1.1 A quick introduction and overview . . . . . . . . . . . . . . . . . . . . . . 1
        1.1.1 Ball-intersection bodies . . . . . . . . . . . . . . . . . . . . . . . 2
        1.1.2 Covering, packing, kissing, and related configurations of balls . . . . . . . 9
    1.2 Background knowledge, notation and definitions . . . . . . . . . . . . . . . . 15

**2 Ball-intersection bodies**   21
    2.1 Ball-hull . . . . . . . . . . . . . . . . . . . . . . . . . . . . . . . . . . . 21
        2.1.1 Definition and basic properties . . . . . . . . . . . . . . . . . . . . 21
        2.1.2 Meissner's bodies . . . . . . . . . . . . . . . . . . . . . . . . . . 26
        2.1.3 Relations between the ball-hull and the ball-intersection of a convex body 29
    2.2 On a theorem of Chakerian . . . . . . . . . . . . . . . . . . . . . . . . . . 33
    2.3 Further characterizations of bodies of constant width . . . . . . . . . . . . . . 36
        2.3.1 Affine orthogonality . . . . . . . . . . . . . . . . . . . . . . . . . 36
        2.3.2 Characterizations of bodies of constant width via affine orthogonality . . 37
        2.3.3 Applications of affine orthogonality for characterizations of further classes of special convex bodies . . . . . . . . . . . . . . . . . . . . . . . 40

**3 Kissing spheres. Coverings and packings by balls**   47
    3.1 Kissing spheres . . . . . . . . . . . . . . . . . . . . . . . . . . . . . . . 47
        3.1.1 Strictly convex, smooth normed planes as topological Möbius planes . . . 48
        3.1.2 Spheres kissing three given spheres . . . . . . . . . . . . . . . . . . 50
    3.2 Covering a disc by translates of the unit disc . . . . . . . . . . . . . . . . . . 54
        3.2.1 Lemmas . . . . . . . . . . . . . . . . . . . . . . . . . . . . . . . 55
        3.2.2 Results . . . . . . . . . . . . . . . . . . . . . . . . . . . . . . . 57
    3.3 Regular 4-coverings . . . . . . . . . . . . . . . . . . . . . . . . . . . . . 60
        3.3.1 Properties of a regular 4-covering . . . . . . . . . . . . . . . . . . . 60
        3.3.2 A lattice covering of the plane based on a regular 4-covering . . . . . . . 65
    3.4 Configurations of circles related to covering problems . . . . . . . . . . . . . 67

|  |  | 3.4.1 | Configurations of Minkowskian circles related to a regular 4-covering | 67 |
|---|---|---|---|---|
|  |  | 3.4.2 | Miquel configurations of circles of equal radii | 68 |
|  |  | 3.4.3 | Miquel configurations of circles having arbitrary radii | 71 |
|  | 3.5 | Visibility in packing of balls | | 75 |
|  |  | 3.5.1 | Special and very special triangles | 76 |
|  |  | 3.5.2 | The concealment number in the planar case | 78 |

**Bibliography** 84

# Chapter 1

# Introduction

## 1.1 A quick introduction and overview

A *normed* (or *Minkowski*) *space* is a finite dimensional linear space equipped with an arbitrary norm. Such spaces are homogeneous (all translations are isometries) but not isotropic. Straight lines are geodesics, and so the study of these spaces falls under the program presented by Hilbert [45] in his fourth problem. The origins and basic developments of the geometry of Minkowski spaces are connected with names like Riemann, Minkowski, and Busemann. More precisely, the earliest contribution to *Minkowski Geometry* was possibly given by Riemann in his "Habilitationsvortrag" [84], where he mentioned the $l_4$-norm. Minkowski [75] introduced the axioms of Minkowski spaces, strongly motivated by relations of this field to the geometry of numbers. Later on, Minkowski geometry was studied by Busemann [26], in order to get a better understanding of *Finsler Geometry* introduced in [38], which is locally Minkowskian; see also [85] and [4]. Closely related is the subject of *Distance Geometry*, going back to Menger [74] and Blumenthal [22]. From a certain point of view, Minkowski Geometry naturally extends results and methods of *Convex and Discrete Geometry*. The present work follows this guideline. I got the related motivation also from participating in writing the surveys [61] and [69] on the geometry of normed spaces. These two surveys shed a light on the geometric aspects of normed spaces. Such an approach is different to that in the classical monograph [99] of Thompson and to the usual approach to normed spaces used in approximation theory and functional analysis. The results in our work show that this different approach is successful in the following sense:

1. We can substantially extend many results from convex and discrete geometry, like properties of bodies of constant width, covering problems, and packing problems.

2. Working in the more general framework of normed space we develop concepts which are completely new, even for the Euclidean subcase. Examples are the notions of ball-hull of a set and affine orthogonality.

3. The derived results are valid in *all normed spaces* or in a large class of them, such as the class of strictly convex normed spaces. It should be noticed that in this more general framework we cannot use the methods usually chosen in the investigations of special norms, as the $l_p$-norm, polyhedral norm, and taxicab norm.

The results in the present work are subdivided into two main topics: ball-intersection bodies and covering/packing/kissing problems.

### 1.1.1 Ball-intersection bodies

We define, in normed spaces, *ball-intersection bodies of size* $\lambda$ as the intersections of (finitely or infinitely numbers of) balls of radius $\lambda$. Until now only special subclasses of the ball-intersection bodies have been intensively studied. Due to the known theorem of Meissner, a body of constant width 1 in Euclidean space is the intersection of all balls of radius 1 centered at this body. According to a theorem of Eggleston, complete bodies in normed spaces have the same property. Another class of ball-intersection bodies that is widely studied is the interesting subclass of bodies of constant width, called Reuleaux triangles in the Euclidean plane as well as in arbitrary normed planes. It should be noticed that Reuleaux polygons are also ball-intersection bodies, but there are not many results on them for non-Euclidean norms. Another interesting appearance of ball-intersection bodies is in an alternative definition of *Jung's constant*. Jung's constant of a normed space $(\mathbb{M}^d, \|\cdot\|)$ is the smallest number such that a ball of diameter being this number may cover, after a suitable translation, any set of diameter $\leq 1$. But it can be also defined as the greatest lower bound on real numbers $\mu$ which possess the following property:

Given any family $\{x_i + \mathcal{B} : i \in I \text{ and } \mathcal{B} \text{ is the unit ball } of (\mathbb{M}^d, \|\cdot\|)\}$ of mutually intersecting balls, then $\cap_{i \in I}(x_i + \mu \mathcal{B}) \neq \emptyset$;

see [42]. Recently another class of ball-intersection bodies became a subject of special interest. This is the class of ball-polyhedra (some authors call them ball-polytopes). A *ball-polyhedron* is defined as the intersection of finitely many balls of the same radius. In a certain sense polytopes can be considered as a special class of ball-polyhedra of size $\infty$. This explains also the use of the term "ball-polytope". For Euclidean space of dimension 3, ball-polyhedra appeared, without using this name, in works of Grünbaum, Heppes, and Straszewicz who gave independent proofs of the *Vázsonyi conjecture*; see, e.g., [81]. Note that the Vázsonyi's conjecture says the maximal number of diameters of a finite set of $n$ points in $\mathbb{R}^3$ is $2n-2$ for $n \geq 4$. Nowadays ball-polyhedra appear again in connection to the Kneser-Poulsen conjecture. A new approach to the Vázsonyi problem inspired researchers very recently to come back to these bodies. It turns out that ball-polyhedra have many interesting properties for themselves, also with respect to some generalized types of convexity, like spindle convexity and ball-convexity. This motivated many researcher

1.1. A QUICK INTRODUCTION AND OVERVIEW

(Capoyleas, Connelly, Csikós, Bezdek, Lángi, Kupitz, Martini, Naszódi, Papez, Perles, etc.) to investigate ball-polyhedra in Euclidean space. Due to the complicated structure of these bodies, most considerations are restricted to dimension 3. For example, the face sructure and the combinatorial structure of ball-polyhedra is completely clarified only in Euclidean space of dimension 3; see [18] and [82]. We present the first systematic approach to ball-intersection bodies in normed spaces. Our investigations are focused on the following directions:

1. We define the *ball-intersection* and the *ball-hull* for a given set (it can be a finite point set or a convex body) in arbitrary normed spaces (they are ball-intersection bodies) and study the relations between these two associated bodies and the original set. The ball-intersections are already known in the literature (e.g., this term is used in the well-known characterization of complete bodies due to Eggleston). The ball-hulls are used until now only in Euclidean space of dimension 2 and 3. In contrast to this we utilize them in any normed space. Applying our methods to a convex body, we get a pair of ball-intersection bodies associated to this body. Although this situation is similar to the one when considering a body and its completion (note that any completion of a convex body is a ball-intersection body), already the study of even two associated bodies by our treatment helps to understand properties of the original convex body in a better and new way.

2. We consider a special class $\mathfrak{C}$ of ball-intersection bodies and we find a subclass $\mathfrak{C}_1$ of it with minimal covering property. More precisely, this means that if a body can be covered by any body from $\mathfrak{C}_1$, then it can be covered by any body from $\mathfrak{C}$. If in the Euclidean plane $\mathfrak{C}$ denotes the family of all bodies of constant width and we replace the minimal covering property by minimal area, then we derive the known theorem of Blaschke-Lebesgue stating that among all bodies of the same constant width Reuleaux triangles have minimal area.

3. We give characterizations of the class of bodies of constant width (forming a special class of ball-intersection bodies) which are new already for the Euclidean subcase. We note that the concept which is developed for these characterizations is also suitable for characterizations of further classes of convex bodies, such as centrally symmetric bodies and ellipsoids.

Now we discuss the concrete results.

⋄ BALL-INTERSECTION AND BALL-HULL

The ball-intersection of a set $M$ in a normed space is defined as the intersection of all balls of the same radii whose centers are from $M$. We introduce the notion of ball-hull of a set $M$ as the intersection of all balls of given radius which contain $M$. In contrast to the notion of ball-intersection, the notion of ball-hull was considered only in the Euclidean subcase as a helpful tool. For instance, Capoyleas [28] and Bezdek, Connelly, and Csikós [16] define the ball-hull of a set in order to give a relation between it and the ball-intersection of the same set. Such a relation is very useful considering different aspects of two basic conjectures of discrete geometry. The first one, known as Kneser-Poulsen conjecture, says that under any contraction of the centers of finitely many balls in a Euclidean space, the volume of the union (respectively, intersection) of these balls cannot increase (respectively, decrease). This conjecture is confirmed only in the planar case by Bezdek and Connelly; see [15]. The second conjecture, namely the conjecture of Alexander, says that under an arbitrary contraction of the center points of finitely many congruent discs in the plane, the perimeter of the intersection of these discs cannot decrease. According to our best knowledge, nothing is known about these two conjectures in normed spaces. Partially, this is due to the lack of investigations of the ball-hull of sets in normed spaces. Note also that Perles, Martini, and Kupitz (see [82] and [58]) considered the ball-hull in Euclidean spaces, in order to give a new approach to the Vázsonyi problem. Now we present the first systematic approach of this notion. We give some basic properties of the ball-hull and investigate the relation of it with the original body and its ball-intersection (our Proposition 2.1.1 and Proposition 2.1.2). We hope (since already there are partial results) that this will be good basis for future investigation in the following direction: it is well known that the relation between a convex body $K$ and its completions (i.e., complete bodies of the same diameter as $K$ which contain $K$) gives a lot of information about $K$. Thus, in our concept, for any convex body $K$ we have three bodies (all the three are ball-intersection bodies) which are associated with them, namely a completion $\mathcal{C}(K)$ of $K$, its ball-intersection $\mathcal{BI}(K)$, and its ball-hull $\mathcal{BH}(K)$. If diam $K$ = diam $\mathcal{BH}(K)$ (this was conjectured for the Euclidean plane by Boltyanski in [23], and it is still open), then the relation

$$K \subseteq \mathcal{BH}(K) \subseteq \mathcal{C}(K) \subseteq \mathcal{BI}(K)$$

holds, i.e., the ball-hull and the ball-intersection of $K$ approximate the completion of $K$. Thus, studying these three bodies, which are associated with a convex body $K$, and the relations between them, we can understand properties of the original convex body in a better and new way.

A further relation between the ball-intersection and the ball-hull of a set $K$ is our Theorem 2.1.8. This theorem holds in all strictly convex, smooth normed planes and says that for any set $K$ of diameter 1 the Minkowski sum of the ball-intersection of $K$ and the ball-hull of $K$ is a convex body of constant width 2. This result extends the results of Capoyleas [28] and Bezdek, Connely, Csikós [16] who proved the same for the Euclidean subcase.

## 1.1. A QUICK INTRODUCTION AND OVERVIEW

Another accent is to investigate when the ball-hull of a convex body coincides with this body. Thus we have an analogue of the theorems of Meissner and Eggleston. As it was mentioned above, according to the theorem of Meissner a convex body of diameter $\lambda$ in a Euclidean space is of constant width if and only if it coincide with its ball-intersection of size $\lambda$. Eggleston proved this for any normed space, but there a convex body is complete if and only if it coincides with its ball-intersection. Note also that going back to Meissner, Kelly, and Eggleston, completeness and constant width are equivalent in Euclidean spaces as well as in two-dimensional normed spaces. In normed spaces of dimension $\geq 3$ constant width implies completeness. Regarding the ball-hull we prove (cf. Theorem 2.1.4) that the class $\mathfrak{C}_1$ of convex bodies coinciding with its ball-hull contains the class $\mathfrak{C}_2$ of convex bodies which coincide with its ball-intersection, i.e., $\mathfrak{C}_2 \subseteq \mathfrak{C}_1$ ($\mathfrak{C}_2$ is the class of complete bodies). In Example 2.1.1 we show that both the classes do not coincide even in the Euclidean subcase. Our Theorem 2.1.5 gives a necessary condition that a body from $\mathfrak{C}_1$ belongs to $\mathfrak{C}_2$.

The ball-intersection of three points forming an equilateral triangle is a Reuleaux triangle. Reuleaux triangles in the Euclidean plane as well as in normed planes are of constant width. But the ball-intersection of four points $p_1, \ldots, p_4$ which form an equilateral tetrahedron in the three-dimensional Euclidean space is no more a body of constant width; see [73]. By flipping couples of circular edges of the ball-intersection of $\{p_1, \ldots, p_4\}$, Meissner constructed a body, usually called *Meissner tetrahedron*, which is of constant width. In [50] Lachand-Robert and Oudet gave an inductive construction of bodies of constant width for Euclidean spaces of arbitrary dimension such that in the three-dimensional case a Meissner tetrahedron is obtained. We apply this construction in normed spaces and prove that the resulting body is complete; see our Theorem 2.1.6. In Euclidean spaces of dimension $\geq 3$ no body of constant width is a ball-polyhedron, i.e., the intersection of finitely many balls of the same radius. But as our Theorem 2.1.7 shows, Meissner tetrahedra can be approximated by ball-polyhedra. More precisely, if we start with the Reuleaux triangle with vertices $p_1, p_2, p_3$, and $p_4$ be a point of the same distance from the points $p_1, p_2, p_3$, then for the resulting Meissner tetrahedron $K$ the following inclusions

$$\mathcal{BH}(p_1, \ldots, p_4) \subset K \subset \mathcal{BI}(p_1, \ldots, p_4)$$

hold, where $\mathcal{BH}(p_1, \ldots, p_4)$ is the ball-hull of $\{p_1, \ldots, p_4\}$ and $\mathcal{BI}(p_1, \ldots, p_4)$ is the ball-intersection of $\{p_1, \ldots, p_4\}$. It should be noticed that this statement is true in a normed space of dimension 3, if this space has the following property:

for any two points $p, q$ belonging to the unit ball $\mathcal{B}$ with $\|p-q\| = 1$ all circular arcs of radius 1 with endpoints $p$ and $q$ also belong to $\mathcal{B}$.

According to our Lemma 2.1.1 the above property holds in any normed plane, but it is no longer true in dimension 3; see the example in Remark 2.1.1.

At the end, it should be noticed that the term "ball-hull" is used by Moreno and Schneider in [77] and [78] in another sense. They define the ball-hull of a set in a normed space as the intersection of all balls containing this set. But very recently (see [76] and [79]) they also use the term "wide spherical hull" of a set $S$. This is, in fact, the ball-hull of S in our sense but for the case when the radii of the balls forming the ball-hull of $S$ are equal to the diameter of S. E.g., in [76] Moreno gives a characterization of the ball-hull of a set.

⋄ A CLASS OF BALL-POLYHEDRA WITH MINIMAL COVERING PROPERTY

One of the most remarkable properties of Reuleaux triangles in the Euclidean plane is that they have minimal area among all bodies of constant width. This result, proved by Lebesque [55] and later by Blaschke [21], is known as the *Blaschke-Lebesque Theorem*. But Reuleaux triangles in the Euclidean plane are also "minimal" in another sense. Namely, if $\mathcal{RT}$ is Reuleaux triangle of width $\lambda$ and any congruent copy $P'$ of a compact, convex set $P$ can be covered by a translate of $\mathcal{RT}$, then $P$ can also be covered by a translate of an arbitrary convex body of constant width $\lambda$. Here a congruent copy $P'$ of a set $P$ means that there exists a translation, or a rotation, or a product of translations and rotations mapping $P$ onto $P'$. This covering property, known as *Chakerian's theorem*, was proved in [31]. Another proof was later given by Bezdek and Connelly; see [14]. Our main result in Section 2.2 presents an extension of Chakerian's theorem to all normed planes. Note that the theorem of Blaschke-Lebesgue was extended to an arbitrary normed plane by Chakerian himself [30] and, independently by Ohmann [80]; see also the survey [70, § 2.8].

⋄ BODIES OF CONSTANT WIDTH AND THE RELATED CONCEPT ON AFFINE ORTHOGONALITY

Section 2.3 is devoted to a new concept of orthogonality for Euclidean as well as for normed spaces. For a convex body $K$ we define an orthogonality relation between two chords $[p_1, p_2]$ and $[q_1, q_2]$ of this body. This relation is not symmetric with respect to both the chords. It is also not symmetric with respect to the endpoints of the first chord, but symmetric with respect to the endpoints of the second chord. For that reason, the notation $[p_1, p_2] \dashv_{p_1} [q_1, q_2]$ is used. Proposition 2.3.1 says that if $[p_1, p_2] \dashv_{p_1} [q_1, q_2]$, then for any chord $[q'_1, q'_2]$ of $K$ that is parallel to $[q_1, q_2]$ the relation $[p_1, p_2] \dashv_{p_1} [q'_1, q'_2]$ holds. Thus, instead of affinely orthogonal chords we can speak about affine orthogonality of a chord and a direction. A similar concept was given by Eggleston. He defined that a chord $[p, q]$ of convex body is a normal of $K$ at $p$ if $[p, q]$ is Birkhoff orthogonal to a supporting hyperplane of $K$ at $p$; see [35, p. 166]). Via the notion of normals of a convex body Eggleston gave a characterization of bodies of constant width

## 1.1. A QUICK INTRODUCTION AND OVERVIEW

which is of interest not only for itself. In the Euclidean subcase this characterization forms the basis for the usual definition of space curves of constant width as well as the generalization of these curves to transnormal manifolds embedded in Euclidean spaces. But our concept of affine orthogonality has the following advantages in comparison to Eglleston's approach.

1. For the definition of affine orthogonality we do not need a metric, i.e., our consideration take place in an arbitrary affine space.

2. We are able to characterize not only bodies of constant width with this notion, but also other classes of convex bodies, such as centrally symmetric ones, ellipses, and bodies whose boundary is a Radon curve.

Note also that the relation of affine orthogonality depends on the considered convex body. If this convex body is a circular disc, then our definition coincides with the usual Euclidean orthogonality. But for bodies of constant width we have coincidence with the notion of normals of Eggleston. Thus the notion of affine orthogonality can be considered on the one hand as a generalization of usual Euclidean orthogonality and, on the other hand, as an extension of Eggleston's concept of normals.

The first characterization in Section 2.3 refers to the notion of normals of Eggleston. Our Theorem 2.3.2 says that a convex body $K$ in a strictly convex and smooth normed plane is of constant width if and only if any normal $[p,q]$ of $K$ at $p$ is affinely orthogonal with respect $p$ to the direction of the supporting line of $K$ at $p$.

In [59] Martini and Makai Jr. gave the following characterization of bodies of constant width in the Euclidean plane $\mathbb{E}^2$: a convex body of diameter 1 in $\mathbb{E}^2$ is of constant width 1 if and only if any two perpendicular chords of it have total length greater or equal to 1. V. Soltan posed the question of extending this characterization to normed planes by replacing usual Euclidean orthogonality by Birkhoff orthogonality. But as the counterexample constructed in [1] shows, in general this cannot be done. We prove that such an extension is possible if Euclidean orthogonality is replaced by affine orthogonality, i.e., in a normed plane a convex body $K$ is of constant width if and only if for any two chords $[p_1, p_2]$ and $[q_1, q_2]$ of $K$, with $[p_1, p_2] \dashv_{p_1} [q_1, q_2]$, the inequality $\|p_1 - p_2\| + \|q_1 - q_2\| \geq \operatorname{diam} K$ holds; see Theorem 2.3.3. Both results, Theorem 2.3.2 and Theorem 2.3.3, are new even in the Euclidean subcase.

As we already mentioned, the relation of affine orthogonality is not symmetric with respect to both the chords, and it is also not symmetric with respect to the endpoints of the first chord. According to our Theorem 2.3.7, the symmetry of the endpoints of the first chord characterize the class of centrally symmetric bodies. Theorem 2.3.10 states that the boundary of centrally symmetric body is a Radon curve if and only if in the relation of affine orthogonality both the chords are symmetric. Theorem 2.3.11 and Theorem 2.3.12 characterize the class of ellipses and the class of circular discs via the notion of affine orthogonality. Moreover, we extend the concept of affine orthogonality to higher dimensions, and Theorem 2.3.5, Theorem 2.3.6, Theorem 2.3.9, and Theorem 2.3.14 give the corresponding characterizations in Euclidean spaces of dimension $\geq 3$.

At the end, we like to explain more detailed the first advantage (see 1. above) of our concept of affine orthogonality, comparing it with Eggleston's concept of normals. We mention two types of non-metrical affine planes, namely the Lorentzian and the isotropic plane. Let $\mathbb{L}^2$ be the vector space $\mathbb{R}^2$ equipped with the *Lorentzian inner product*

$$x \cdot y = x_1 y_1 - x_2 y_2 \quad \text{for} \quad x = (x_1, x_2), \; y = (y_1, y_2).$$

The affine plane associated to the Lorentzian vector space $\mathbb{L}^2$ is called the *Lorentzian plane*; see, e.g., [20] and [102, § 11 and § 12]. It should be noticed that the terms "pseudo-Euclidean" or "Minkowski" are also used for this plane. Each of the two parts of the curve

$$(x_1 - p_1)^2 - (x_2 - p_2)^2 = \lambda^2$$

is called a (timelike) *Lorentzian circle*. From the viewpoint of Klein's concept of geometry the absolute of the Lorentzian geometry consists of two points, for example $f_1 = (1, 1, 0)$ and $f_2 = (1, -1, 0)$ with respect to an affine coordinate system with homogeneous coordinates, such that the line at infinity has the equation $x_3 = 0$. In a certain way (adding points at infinity) we can consider a Lorentzian circle as a closed convex curve. Then two chords are affinely orthogonal if and only if the corresponding vectors are Lorentzian orthogonal, i.e., its Lorentzian inner product is 0; for more details see [3, §5]. The *isotropic plane* is defined as a projective plane with absolute (in the sense of Klein) consisting of a line $F$ and a point $f$ on this line; see, e.g, [86] and [102, Chapter 1 and Chapter 2]. This plane is also called Galilean plane, since its group of motions describes Galileo's principle of relativity. This principle says that all properties studied in mechanics are preserved under transformations of the physical system obtained by imparting to it a velocity which is constant in magnitude and direction, i.e., under so-called Galilean transformations. An *isotropic circle* is a conic touching $F$ at $f$. Any line through $f$ is *isotropic orthogonal* to an arbitrary line; see [86]. Again we can consider an isotropic circle as a convex closed curve, and then all affine diameters are the chords through $f$. Thus we have that two chords $[p_1, p_2]$, $[q_1, q_2]$ of an isotropic circle are affinely orthogonal if and only if they are isotropic orthogonal. These two examples show that the concept of affine orthogonality generalizes the concepts of Lorentzian and isotropic orthogonality in the corresponding planes and can be applied in a more general sense than this one, which is considered in our Section 2.3.

## 1.1.2 Covering, packing, kissing, and related configurations of balls

Common methods of Minkowski geometry are also widely used for solving covering and packing problems. This is due to the fact that any symmetric body can be viewed as a ball with respect to some norm. Thus the corresponding problem can be described and solved in terms of discs (with the same or different radii) in normed spaces. It is possible to use this approach even for non-centrally symmetric bodies. Then the unit ball (in this situation is called a *gauge*) is not assumed to be centrally symmetric, but it can also induce a "norm", of course without the symmetry property. In fact, Minkowski, who introduced the norm axioms, did not assume symmetry. Another way is to consider the *central symmetral* for a convex body $K$, i.e., $K^* = \frac{1}{2}(K - K)$, which is centrally symmetric. Then, as it has been noted by Minkowski and used also by Hadwiger and Grünbaum (see [44] and [43]), $(x+K) \cap (y+K)$ and $(x+K^*) \cap (y+K^*)$ are simultaneously empty, non-empty, or have interior point.

⋄ Kissing spheres

The concept of kissing is more general than that of non-overlapping. In a normed space two balls $B_1$ and $B_2$ do not overlap if they intersect, but do not have common interior points. Let now $B_1$ and $B_2$ be two balls, and $S_i$, $i = 1, 2$, be the boundary of $B_i$. The spheres $S_1$ and $S_2$ are called *kissing spheres* if $S_1 \cap S_2 \neq \emptyset$ and one of the following situations holds:

1. $B_1$ and $B_2$ are non-overlapping balls;

2. $B_1 \subsetneq B_2$;

3. $B_2 \subsetneq B_1$.

Suitably we define also kissing hyperplanes. Let $\mathfrak{F}$ be a family of spheres and hyperplanes. It is said to be a *kissing family* if any two members of this family are kissing each other. Our results in Section 3.1 can be grouped in the following directions:

1. For a given kissing 3-member family of circles and lines in a strictly convex normed plane we find at least one or even all circles that kiss all members of this family.

2. In a strictly convex and smooth normed plane we find *all* circles kissing any member of a given 3-member family of circles and lines which are in arbitrary position to each other. Here to the given family we include also points, interpreted as circles of radius 0.

The used methods in both the cases are quite different. For the first case we study the behavior of curves topologically equivalent to a line. For the second case we prove, in Theorem 3.1.1, that any strictly convex, smooth plane is also topological Möbius plane, i.e., an incidence structure of points and circles satisfying some incidence axioms, and points and circles carry topologies such that the geometric operations are continuous. This interpretation of strictly convex, normed planes is completely new and can also be used for clarifying other types of problems in normed planes.

Referring to 1., we have the following results. Theorem 3.1.5, which asserts that for any kissing family of three circles of the same radii there exist exactly two circles kissing the given circles, holds in any strictly convex normed plane. To prove this theorem we need the following fact, that is of interest for itself and is also used for solving some covering problems. For three non-collinear points in a strictly convex normed plane not always a circle exists containing them. But if these three points form an equilateral triangle then, according to our Lemma 3.1.2, such a circle exists. The next result, namely Theorem 3.1.6, also holds in any strictly convex normed plane. It says that for any given family of two circles touching each other externally and their common supporting line always a circle exists kissing all members of this family.

Our results referring to 2. hold in arbitrary strictly convex and smooth normed planes. According to Theorem 3.1.2, for any generalized circle $K$ (i.e., a circle or a line) and two points $p_1, p_2 \notin K$, which are not separated by $K$, there exist exactly two generalized circles through $p_1$ and $p_2$ kissing $K$. Theorem 3.1.3 gives, for an arbitrary family of two circles and a point, the exact number of generalized circles kissing all member of this family. Let now $K_1, K_2, K_3$ be three pairwise intersecting generalized circles (i.e., properly intersecting or kissing) in a normed plane $(\mathbb{M}^2, \|\cdot\|)$. A set $T$ is called a *circular triangle* if it is a connected component of $\mathbb{M}^2 \setminus (\cup_{i=1}^3 K_i)$ such that each $K_i \cap \text{bd } T$ is connected and has nonempty interior in $K_i$. In Theorem 3.1.4 it is proved that for any circular triangle $T$ formed by $K_1, K_2, K_3$ there exists precisely one generalized circle $K$ kissing $K_1, K_2, K_3$, which belongs to the closure of $T$.

⋄ Covering problems in normed planes

The problem of covering the Euclidean unit disc with $k$ homothets of it having minimum diameter is called the *circle*[1] *covering problem* (see, e.g., [37] and [24] for the relevant results). In Section 3.2 we investigate the extension of this circle covering problem to normed planes for $k \in \{3, 4\}$. Our considerations can also be viewed as contributions to the following more general problem. Let $K$ be a convex body and denote by $h_k(K)$ the smallest positive ratio of $k$ homothetical copies of $K$ whose union covers $K$. For $k \in \{3, 4\}$ the following bounds on $h_k(K)$ are known :

$$\frac{2}{3} \leq h_3(K) \leq 1, \tag{1.1}$$

---

[1]The term "circle" here is not correct, but so is the tradition in the literature regarding covering and packing problems.

## 1.1. A QUICK INTRODUCTION AND OVERVIEW

$$\frac{1}{2} \leq h_4(K) \leq \frac{\sqrt{2}}{2}; \tag{1.2}$$

see [54]. According to [54], [37], and my best knowledge for $k \in \{5, 6, 7, 8\}$ the exact lower bounds on $h_k$ are not known, and for $k \in \{5, 6\}$ the exact upper bounds. In this context we give an exact geometric description of $h_k(K)$ for $k = \{3, 4\}$ in terms of the radius of inscribed equilateral polygons if $K$ is a strictly convex, centrally symmetric convex body. Note that an optimal *packing* of a minimal homothecical copy of the unit disc by unit discs is described by Doyle, Lagarias, and Randall (see [34]) in such terms. It should be noticed that the case $k = 7$ for a centrally symmetric body was investigated by Lassak; see again [54]. In order to give such a geometric description, we reformulate the above problem for a centrally symmetric convex body $K$, i.e., $K$ can be considered as the unit disc $\mathcal{D}$ with respect to some norm. Let $R_k(\mathcal{D})$ be the maximal radius of all homothets of $\mathcal{D}$ that can be covered by $k$ translates of $\mathcal{D}$. Then

$$R_k(\mathcal{D}) = \frac{1}{h_k(\mathcal{D})},$$

and one can rewrite the inequalities (1.1), (1.2) as follows:

$$1 \leq R_3(\mathcal{D}) \leq \frac{3}{2}, \tag{1.3}$$

$$\sqrt{2} \leq R_4(\mathcal{D}) \leq 2.$$

It should be also noticed that the fact whether a planar convex body can be covered by 3 or 4 smaller positive homothetical copies of itself is not apriori true. For 3 homothets this was proved by Levi [56] which confirmed for the planar case the famous conjecture of Hadwiger saying that any convex body of the $d$-dimensional Euclidean space can be covered by $2^d$ smaller homothetical copies of itself.

Now we announce the related theorems which give the exact geometric description of $R_3(\mathcal{D})$ and $R_4(\mathcal{D})$, where $\mathcal{D}$ is the unit disc of a strictly convex normed plane. Theorem 3.2.1 says that the circumdisc of any equilateral triangle of side-length 2 has radius > 1 and can be covered by three translates of the unit disc. Moreover, according to our Theorem 3.2.2 the quantity $R_3(\mathcal{D})$ is the maximal circumradius of equilateral triangles with side-length 2. Proposition 3.2.1 gives an upper bound on $R_3(\mathcal{D})$, namely $R_3(\mathcal{D}) \leq \frac{4}{3}$. This upper bound strengthens the second inequality in (1.3) for the case that $\mathcal{D}$ is centrally symmetric and strictly convex. If bd $\mathcal{D}$ is an affine regular hexagon, then $R_3(\mathcal{D}) = \frac{4}{3}$. For the case of four covering discs we prove that the quantity $R_4(\mathcal{D})$ is the maximal circumradius of all parallelograms whose four sides are of Minkowskian length 2, and whose two diagonals have the same length; see Theorem 3.2.4. This result refers to the problem of covering a convex body in the plane by four smaller homothets of it, i.e., the planar case of the conjecture of Hadwiger mentioned above. The smallest possible ratio of those four homothets is attained at the so-called *regular 4-covering*. This regular 4-covering was constructed by Lassak [53] in order to prove that the smallest possible ratio $\lambda$ is $\leq \frac{\sqrt{2}}{2}$; see (1.2). In Section 3.3 we continue the investigations of Lassak on regular 4-coverings and derive further properties of such coverings. We prove that for any convex body $K$ of diameter 1 and a regular 4-covering of it the smallest homothetical copy of $K$, which contains this regular 4-covering, is of diameter $2\lambda$. Note that this result (our Theorem 3.3.1) refers to *all* convex bodies. But if the considered body $K$ is centrally symmetric, Theorem 3.3.2 gives a detailed description how the discs of the covering are placed to each other and with respect to $K$. Based on a regular 4-covering of a centrally symmetric convex body, we construct a lattice covering of the plane. This covering has margin zero and induces a packing of the plane (Proposition 3.3.1). We investigate also the *Voronoi region* and the *gray area* of an element of this lattice covering. We only mention two of the properties of them. Namely, the family of translates of a Voronoi region, obtained by the basis vectors of the lattice, is a tiling of the plane. The convexity of a Voronoi region implies that it is a parallelogram. The other properties are summarized in Proposition 3.3.2.

⋄ CONFIGURATIONS OF CIRCLES RELATED TO COVERINGS

We return to the configuration of circles in a strictly convex normed plane described in Theorem 3.3.2. It consists of four circles $C_i$, $i = 1, \ldots, 4$, of the same radius $\lambda$ passing through a point $p$ such that $C_i$ and $C_{i+1}$ do not touch each other. Then $C_i$ and $C_{i+1}$ have exactly one second intersection point, denoted by $p_{i+1}$. If this configuration is a regular 4-covering of a disc $D$ with center $p$ and radius $\mu$, where $\mu > \lambda$, then the points $p_1, \ldots, p_4$ lie on the same circle of radius $\mu$. In case of *three* circles of the same radius passing through a point $p$, the second intersection points (if any two of the circles do not touch) always lie on a circle of the same radius. In the Euclidean case this is the so-called Țițeica theorem. This theorem was extended by Asplund and Grünbaum [6] to strictly convex, smooth normed planes, but it also holds if these planes are not necessarily smooth; see [60] and also [62] for further extensions. Our Theorem 3.4.1 clarifies what configuration is obtained if the second intersection points of even *four* circles of the same radius passing through a point $p$ also lie on one circle. More precisely, in a strictly convex normed plane, let there be given four circles $C_i$, $i = 1, \ldots, 4$, of radius $\lambda$ passing though a point $p$ such that $C_i$ and $C_{i+1}$ do not touch each other, whereas $C_i$ and $C_{i+2}$ touch each other. If $p_{i+1}$ ($p_5 \equiv p_1$) is the second intersection point of $C_i$ and $C_{i+1}$ and $p_1, p_2, p_3, p_4$ lie on the same circle of radius $\mu > \lambda$, then the union of the discs having these four circles as boundaries is a regular 4-covering of the disc, with center $p$ and radius $\mu$.

More general than the above configurations is a configuration $\{C_i, i = 1, \ldots, 4\}$, where the first intersection points of $C_i$ and $C_{i+1}$ do not coincide. We can describe also this configuration

## 1.1. A QUICK INTRODUCTION AND OVERVIEW

in the following way. Let $p_1, \ldots, p_8$ be eight points. To every point $p_i$, $i = 1, \ldots, 8$, we assign a vertex of a cube. Consider the six quadruples of points that correspond to the vertices of each facet of the cube, e.g.,

$$(p_1, p_2, p_3, p_4), (p_1, p_2, p_5, p_6), (p_2, p_3, p_7, p_6),$$
$$(p_3, p_4, p_8, p_7), (p_1, p_4, p_8, p_5), (p_5, p_6, p_7, p_8). \tag{1.4}$$

If five of the quadruples in (1.4) are concyclic (i.e., there exists a circle passing through all points of the quadruple), then this configuration is called a *Miquel configuration*. In the Euclidean plane *all six* quadruples are concyclic. This statements is known as *Miquel's Theorem*. It was proved by Asplund and Grünbaum in [6] that in a strictly convex and smooth normed plane, for any Miquel configuration of circles with equal radii the theorem of Miquel holds. Our Theorem 3.4.3 extends the result of Asplund and Grünbaum to *all* normed planes. To answer the question what happens in a Miquel configuration of circles of arbitrary radii in normed planes we return to Theorem 3.1.1 which says that every strictly convex, smooth normed plane is a Möbius plane. Let $\Sigma$ and $\Sigma'$ be two Möbius planes. If there exists a one-to-one correspondence $\sigma : \Sigma \to \Sigma'$ mapping concyclic points into concyclic points, and non-concyclic points into non-concyclic ones, then $\Sigma$ and $\Sigma'$ are called *isomorphic* and $\sigma$ is said to be a *homography from $\Sigma$ to $\Sigma'$*. Let now $(\mathbb{M}^2, \|\cdot\|)$ be a strictly convex, smooth normed plane, and consider this plane as a Möbius plane $\Sigma = (\mathfrak{P}, \mathfrak{C})$. A homography $\varphi$ in $\Sigma$ that is involutory and leaves the points of a circle $C$ fixed such that no other point is fixed is called *the inversion with respect to the circle $C$*. It is clear that such a homography exists at least for the Euclidean subcase. We prove in Theorem 3.4.5 that in a strictly convex, smooth normed plane there exists an inversion $\varphi$ with respect to some circle of $(\mathbb{M}^2, \|\cdot\|)$ if and only if the plane is Euclidean. In [96], Stiles gives another definition for inversion. Namely, he defines the inversion with respect to the unit circle $\mathcal{C}$ of a normed plane $(\mathbb{M}^2, \|\cdot\|)$ as a mapping $\varphi$ of $\mathbb{M}^2 \setminus \{0\}$ onto itself that maps a point $x \neq 0$ onto the point $\frac{1}{\|x\|^2} x$. He proves that if the inversive image of some line is a circle, then $(\mathbb{M}^2, \|\cdot\|)$ is Euclidean. Theorem 3.4.5 shows that Stiles' definition of inversion and ours are only equivalent in the Euclidean case. As a consequence of Theorem 3.4.5 we have also that if Miquel's theorem holds in a strictly convex, smooth normed plane $(\mathbb{M}^2, \|\cdot\|)$, then this plane is Euclidean (Theorem 3.4.6). The proof of Theorem 3.4.6 shows that the condition that $(\mathbb{M}^2, \|\cdot\|)$ is a Miquelian Möbius plane can be replaced by the condition that $(\mathbb{M}^2, \|\cdot\|)$ is isomorphic to a Möbius plane $\Sigma' = \text{Mo}(\mathbb{F}, \mathbb{E})$, where $\mathbb{F}$ is a commutative field and $\mathbb{E}$ is a quadratic extension of $\mathbb{F}$. Now we explain the construction of $\Sigma'$. One can consider the elements of $\mathbb{E} \cup \{\infty\}$, where $\infty$ is a formal symbol, as points and define circles (usually called *chains*) as sets

$$\{x \in \mathbb{E} \cup \{\infty\} | \frac{p-r}{q-r} : \frac{p-x}{q-x} \in \mathbb{F} \cup \infty\},$$

where $p, q$, and $r$ are three pairwise different points from $\mathbb{E}$. Then the so-defined incidence structure is a Möbius plane (see [11, § 2]), and we denote it by $\mathrm{Mo}(\mathbb{F}, \mathbb{E})$. Note also that in the classical case $\mathbb{F} = \mathbb{R}$, $\mathbb{E} = \mathbb{C}$, and $\mathrm{Mo}(\mathbb{F}, \mathbb{E})$ is the inversive plane. Thus Theorem 3.4.6 shows that non-Euclidean, strictly convex, smooth normed planes cannot be algebraized. Note that Theorem 3.4.6 is not only a characterization of the Euclidean plane among all strictly convex, smooth planes. This theorem also clarifies the place of strictly convex, smooth normed planes among all Möbius planes, i.e., non-Euclidean strictly convex, smooth normed planes are non-Miquelian Möbius planes.

### ⋄ Concealment number of normed space

The last section is devoted to visibility in a packing of translates of the unit disc of normed plane. Let $\mathfrak{F}$ be a family of translates of $K$ packed between disjoint translates $p + K$ and $q + K$. Then $p + K$ and $q + K$ are called *visible* from each other in the packing $\{p + K, q + K\} \cup \mathfrak{F}$ if there exist points $x \in (p + K)$ and $y \in (q + K)$ such that the segment $[x, y]$ intersects no element of $\mathfrak{F}$. Otherwise $p + K$ and $q + K$ are said to be *concealed* from each other by $\mathfrak{F}$. If $\mathcal{B}$ denotes the unit ball of a normed plane $(\mathbb{M}^d, \|\cdot\|)$, the *concealment number* $\delta((\mathbb{M}^d, \|\cdot\|))$ is defined as the infimum of $\lambda > 0$ satisfying the following condition: for $p + \mathcal{B}$ and $q + \mathcal{B}$ being disjoint, the inequality $\|p - q\| > \lambda$ implies that $p + \mathcal{B}$ and $q + \mathcal{B}$ can be concealed from each other by packing translates of $\mathcal{B}$ between them. For any norm we have the inequality

$$\delta((\mathbb{M}^d, \|\cdot\|)) \leq 4.$$

It is also easy to check that for the Euclidean plane $\mathbb{E}^2$ the equation $\delta(\mathbb{E}^2) = 2\sqrt{3}$ holds. In Section 3.5 we extend the investigations of Hosono, Maehara, and Matsuda ([48]) on concealment numbers from the Euclidean case to two-dimensional normed spaces. We define the *concealment number $\delta_p$ of the direction $p$*. It is the infimum of $\mu > 2$ such that the unit disc $\mathcal{D}$ and its translate $\mu p + \mathcal{D}$ can be concealed from each other by packing translates of $\mathcal{D}$. Clearly, $\delta_p = \delta_{-p}$ and $\delta(\mathbb{M}^2, \|\cdot\|) = \sup\{\delta_p : p \in \mathcal{C}\}$. To describe the concealment number we introduce the notion of special and very special triangles. These notions are of interest of themselves, since in terms of them one can get characterizations of the Euclidean plane among all normed planes; see [2, Theorem 3.1 and Corrolary 3.1] Our main result in Section 3.5, namely Theorem 3.5.1, says:

(i) If $[p, q]$ is the base of a special triangle, then $\delta_{\frac{p-q}{\|p-q\|}} \geq \|p - q\|$.

(ii) If $[p, q]$ is the base of a very special triangle, then $\delta_{\frac{p-q}{\|p-q\|}} = \|p - q\|$.

(iii) If $\|p - q\| < 4$ and $[p, q]$ is the base of a special triangle that is not very special, then $\delta_{\frac{p-q}{\|p-q\|}} > \|p - q\|$.

For strictly convex normed planes, Busemann and Kelly defined reflections in a line as isometries having this line as line of fixed points; see [27, p. 127]. Not every line admits a reflection in itself; cf. [27, p. 140, Theorem 25.3]. According to our Corollary 3.5.1 we have that if the segment $[p, q]$ is the base of a special triangle with $\|p - q\| < 4$ and the line through $p$ and $q$ admits a reflection, then

$$\delta_{\frac{p-q}{\|p-q\|}} = \|p - q\|.$$

Note that a different approach to reflections in normed planes can be found in [67] and [68]. In contrast to the approach of Busemann and Kelly the reflections in lines there are defined as affine transformations that are not necessarily isometries.

## 1.2 Background knowledge, notation and definitions

Let $\mathbb{M}^d$ be a finite dimensional real vector space. In the sequel the elements of $\mathbb{M}^d$ will be denoted by $x, y, \ldots$, and the origin by $0$. The elements of $\mathbb{M}^d$ will often be interpreted as points in a real affine space. If $p_1, \ldots, p_d$ are affinely independent points, we denote the *hyperplane* that is the affine hull of $p_1, \ldots, p_d$ by $\mathrm{HP}(p_1, \ldots, p_d)$. For the *line* (or *1-flat*) through the different points $p$ and $q$ we simply write $\mathrm{L}(p, q)$. We denote the closed half space bounded by $\mathrm{HP}(p_0, \ldots, p_{d-1})$ and containing the point $p \notin \mathrm{HP}(p_0, \ldots, p_{d-1})$ by $\mathrm{HS}_p^+(p_0, \ldots, p_{d-1})$, and by $\mathrm{HS}_p^-(p_0, \ldots, p_{d-1})$ this one not containing $p$. If the hyperplane $G$ is a translate of the hyperplane $H$, then for the half space, which is bounded by $G$ and contains $H$, we write $\mathrm{HS}_H^+(G)$. A ray emanating from $p$ and containing $x$ is denoted by $\mathrm{R}_x^+(p)$. We use the denotation $\mathrm{R}_x^-(p)$ for the ray opposite to $\mathrm{R}_x^+(p)$. As usual, the *convex hull* of a set $K \subset \mathbb{M}^d$ is denoted by $\mathrm{conv}\, K$. If $p, q, r$ are three non-collinear points, by a *triangle* with vertices $p, q, r$ we mean the convex hull of $\{p, q, r\}$. If the points $p, q, r$, and $s$ are such that any of them does not belong to the convex hull of the other three, then the convex hull $\mathrm{conv}\,\{p, q, r, s\}$ is a *convex quadrangle* with vertices $p, q, r$, and $s$. Through this work, the notations $\mathcal{T}(p, q, r)$ and $\mathcal{Q}(p, q, r, s)$ will be used for the triangle with vertices $p, q, r$ and for the quadrangle with vertices $p, q, r, s$, respectively. Note that, although this geometric interpretation, the elements of $\mathbb{M}^d$ can also be matrices, polynomials of degree less than $d$, etc. Defining an inner product in $\mathbb{M}^d$ one obtains a *Euclidean space* denoted by $\mathbb{E}^d$. A natural generalization of the Euclidean space is the notion of *normed (or Minkowski) space*. It is defined as a finite dimensional real vector space $\mathbb{M}^d$ equipped with a real valued function $\|\cdot\| : \mathbb{M}^d \longrightarrow \mathbb{R}$, called its *norm*, which satisfies the following conditions:

- $\|x\| \geq 0$ and $\|x\| = 0 \iff x = 0$;
- $\|\lambda\, x\| = |\lambda|\, \|x\|$, where $\lambda \in \mathbb{R}$;
- $\|x + y\| \leq \|x\| + \|y\|$.

Let $(\mathbb{M}^d, \|\cdot\|)$ be a normed space with *unit ball* $\mathcal{B} := \{x \in \mathbb{M}^d : \|x\| \leq 1\}$ and *unit sphere* $\mathcal{S} := \{x \in \mathbb{M}^2 : \|x\| = 1\}$. A homothetical copy $\lambda \mathcal{B} + p$ of $\mathcal{B}$, where $p \in \mathbb{M}^d$ and $\lambda \in \mathbb{R}^+$, is called the (Minkowskian) *ball with center $p$ and radius $\lambda$*. We denote $\lambda \mathcal{B} + p$ by $\mathcal{B}(p, \lambda)$. Analogously, $\lambda \mathcal{S} + p$ is the (Minkowskian) *sphere with radius $\lambda$ centered at $p$* and denoted by $\mathcal{S}(p, \lambda)$. Sometimes, if there is no possibility of misunderstanding, we say a unit ball (unit sphere) for a ball (sphere) of radius 1, i.e., for a translate of the unit ball (sphere). A ball and a sphere in a two-dimensional normed space are called a *disc* and a *circle*, respectively. Let $\lambda > 0$, and $p, q$ be points of a normed plane $(\mathbb{M}^2, \|\cdot\|)$ with $\|p - q\| < 2\lambda$. Then the circles $\mathcal{C}(p, \lambda)$ and $\mathcal{C}(q, \lambda)$ intersect; see, e.g., [72, Lemma 13]. If $x$ is a point such that $\|p - x\| = \lambda$ and $\|q - x\| = \lambda$, then the *circular arc of radius $\lambda$ with center $x$ joining $p$ and $q$* is defined to be the set

$$\{x + \alpha(p - x) + \beta(q - x) : \alpha, \beta \geq 0 \text{ and } \|\alpha(p - x) + \beta(q - x)\| = \lambda\}.$$

We denote it by $\mathrm{arc}_\lambda(p, q; x)$. If the radius $\lambda$ is 1, we simply write $\mathrm{arc}(p, q; x)$.

A normed space $(\mathbb{M}^d, \|\cdot\|)$ is called *strictly convex* if the equality $\|x + y\| = \|x\| + \|y\|$ implies that $x$ and $y$ are linearly dependent. Geometrically this means that the unit sphere does not contain a non-trivial line segment. A normed space is said to be *smooth* if the norm is differentiable at each non-zero point, or equivalently if the unit ball has a unique supporting hyperplane at each boundary point.

Any norm accomplishes $\mathbb{M}^d$ with a topological structure and we use the standard denotations int $K$, cl $K$, and bd $K$ for the *interior*, the *closure*, and the *boundary* of a set $K$ from $\mathbb{M}^d$, respectively.

A non-zero vector $p \in \mathbb{M}^d$ is *Birkhoff orthogonal* to a non-zero vector $q \in \mathbb{M}^d$, denoted by $p \dashv q$, if for any real $\lambda$ the inequality $\|p\| \leq \|p + \lambda q\|$ holds. This means that there is a supporting line of $\|p\|\mathcal{S}$ at $p$ being parallel to $q$, where $\mathcal{S}$ is the unit sphere of $(\mathbb{M}^d, \|\cdot\|)$. Clearly, the so-defined relation of orthogonality is not symmetric in general. For $d \geq 3$ the symmetry of Birkhoff orthogonality implies that the space $(\mathbb{M}^d, \|\cdot\|)$ is Euclidean; see, e.g., [19]. But in the two-dimensional case, according to the Busemann theorem ([25], see also [71, Theorem 1]) for every norm $\|\cdot\|$ there is a norm $\|\cdot\|_a$, unique up to a factor and called *antinorm* of $\|\cdot\|$, such that for any $x, y$ with $x \dashv y$ we have that $y$ is normal to $x$ with respect to $\|\cdot\|_a$, denoted by $y \dashv_a x$, i.e., for any $\lambda \in \mathbb{R}$ the inequality $\|x\|_a \leq \|x + \lambda y\|_a$ holds. This antinorm of $\|\cdot\|$ can be defined in the following way. Let $(\mathbb{M}^2)^*$ be the dual plane of $\mathbb{M}^2$, i.e., $(\mathbb{M}^2)^*$ is the two-dimensional vector space consisting of all linear transformations $\varphi : \mathbb{M}^2 \to \mathbb{R}$. On $(\mathbb{M}^2)^*$ a norm $\|\cdot\|^*$ can be introduced by

$$\|\varphi\|^* := \sup\{\varphi(x) : \|x\| = 1\}.$$

## 1.2. BACKGROUND KNOWLEDGE, NOTATION AND DEFINITIONS

The dual plane $(\mathbb{M}^2)^\star$ can be identified with $\mathbb{M}^2$. This identification can also be realized by an isomorphism

$$\tau : \begin{cases} \mathbb{M}^2 \to (\mathbb{M}^2)^\star \\ \tau(y) \longmapsto [\cdot, y], \end{cases}$$

where $[\cdot, \cdot]$ is a symplectic bilinear form. Thus one can define a new norm on $\mathbb{M}^2$, namely an antinorm, by

$$\|x\|_a := \|\tau(x)\|^\star.$$

For detailed considerations on the antinorm we refer to [71]. Here we only mention that the antinorm of the antinorm is the original norm (see, e.g., [71, Proposition 1]), which together with the Buseman theorem implies that

$$x \dashv y \iff y \dashv_a x \tag{1.5}$$

The unit circle of $\|\cdot\|_a$ is called *unit anticircle* with respect to $\|\cdot\|$ and denoted by $\mathcal{C}_a$. If $\mathcal{C}_a$ is a homothetic copy of $\mathcal{C}$, then, of course, the normality relation is symmetric, and planes with this property are called *Radon planes*.

Since the so-defined relation of Birkoff orthogonality is homogeneous, in what follows we also use the term of Birkoff orthogonal directions. We note also that in a two-dimensional, strictly convex normed space for any direction there exists exactly one direction Birkhoff orthogonal to it. In a two-dimensional smooth normed space there exists, for any given direction, exactly one direction such that the given direction is normal to this direction.

In a natural way we also may define the relation of Birkoff orthogonality for a vector and a hyperplane. A non-zero vector $p$ is said to be Birkoff orthogonal to a hyperplane $G$, if there exists a hyperplane $G'$ parallel to $G$ which supports $\|p\|\mathcal{S}$ at $p$.

If $G_1$ and $G_2$ are two parallel hyperplanes, then the *distance* $\delta(G_1, G_2)$ between them is defied by

$$\delta(G_1, G_2) := \inf\{\|x_1 - x_2\| : x_1 \in G_1, x_2 \in G_2\}.$$

This infimum is only attained at points $x_1 \in G_1$, $x_2 \in G_2$ for which $\mathrm{L}(x_1, x_2) \dashv G_1, G_2$ holds. Analogously, $\delta(p, G) := \inf\{\|p - q\| : q \in G\}$ is the distance between a point $p$ and a hyperplane $G$.

The *bisector* of two points $p$ and $q$ in a normed space $(\mathbb{M}^d, \|\cdot\|)$ is defined by

$$B(p, q) := \{x \in \mathbb{M}^d : \|x - p\| = \|x - q\|\}.$$

In general, bisectors have very complicated topological structure even in dimension 3, but in every strictly convex normed plane they are unbounded simple curves; see [71, § 8.2] and the survey [70, §4.2]. Also we note that the definition of bisectors immediately implies that $B(p, q)$

is symmetric with respect to the midpoint of $[p, q]$. Many further results on and applications of bisectors in normed planes and spaces are collected in Part 4 of the survey [70].

The first lemma which is necessary for our considerations is known as the *monotonicity lemma*. It is proved in [41]; see also [99, Lemma 4.1.2] and [72, § 3.5].

**Lemma 1.2.1.** *Let $\mathcal{C}$ be the unit circle in a normed plane $(\mathbb{M}^2, \|\cdot\|)$, and $p, q, r$ be different points belonging to $\mathcal{C}$ such that the origin $0$ does not belong to the open half-plane determined by $\mathrm{L}(p, q)$ which contains $r$. Then*

$$\|p - q\| \geq \|p - r\|,$$

*with equality if and only if $q, r$, and $\frac{1}{\|q-p\|}(q - p)$ belong to a segment contained in $\mathcal{C}$.*

The next lemmas express basic metrical relations in different point configurations. The first one is a special case of Corollary 28 in [72].

**Lemma 1.2.2.** *Let $p, q, x$ be three non-collinear points in a normed plane $(\mathbb{M}^2, \|\cdot\|)$ and $y \in \mathrm{int}\,\mathrm{conv}\,\{p, q, x\}$. Then $\|p - y\| + \|y - q\| < \|p - x\| + \|x - y\|$.*

**Lemma 1.2.3.** ([72, Lemma 5]) *In a normed space $(\mathbb{M}^d, \|\cdot\|)$, let there be given two distinct points $p$ and $q$ and a point $x$ which is strictly between $p$ and $q$. For an arbitrary point $y$ of $\mathbb{M}^d$ the inequality $\|y - x\| \leq \max\{\|y - p\|, \|y - q\|\}$ holds. Equality is possible if and only if $\|y - x\| = \|y - p\| = \|y - q\|$.*

**Lemma 1.2.4.** ([72, Proposition 7]) *If the points $p, q, r, s$ form a convex quadrangle in a normed plane, then the sum of its diagonals is at least the sum of two opposite sides, i.e.,*

$$\|p - r\| + \|q - s\| \geq \|p - q\| + \|s - r\| \text{ and } \|p - r\| + \|q - s\| \geq \|p - s\| + \|q - r\|.$$

Now we discuss the intersection of two circles in a normed plane. In general, this intersection is always the union of two segments, which are either disjoint or intersect in a common point. Each of these segments may degenerate to a point or to the empty set. This was proved by Grünbaum [41] and later also by Banasiak [8]. The next lemma, referred to as Proposition 21 in [72], describes the intersection precisely. Moreover, it includes a statement on where the different pieces of the circles lie relative to each other, which is a generalization of a lemma of Schäffer; see [91, Lemma 4.3].

**Lemma 1.2.5.** *Let $C$ and $C'$ be two circles in a normed plane $(\mathbb{M}^2, \|\cdot\|)$. Then $C \cap C'$ is a union of two segments $A_1$ and $A_2$, each of which may degenerate to a point or to the empty set. Let the point $p_i \in A_i$, $i = 1, 2$, and let $\varphi : C \to C'$ be the positive homothety which maps*

## 1.2. BACKGROUND KNOWLEDGE, NOTATION AND DEFINITIONS

$C$ into $C'$ (or translation, if $C$ and $C'$ are of the same radii). Let $c_i = \varphi^{-1}(p_i)$ and $c'_i = \varphi(p_i)$. Let $\gamma_1$ ($\gamma_2$) be the part of $C$ on the same side (opposite side) of $L(p_1, p_2)$ as $c_1$ and $c_2$; similarly for $\gamma'$. Then $\gamma_2 \subseteq \operatorname{conv} \gamma'_1$ and $\gamma'_2 \subseteq \operatorname{conv} \gamma_1$.

If $(\mathbb{M}^2, \|\cdot\|)$ is strictly convex, then the intersection of two circles described in the above lemma is very simply. According to [72, Proposition 14], if it is not empty, it only consists of one or two points. If two spheres in normed space have exactly one common point we say that they *touch* each other. The next lemma is usually used without proof. Its proof is really trivial and we omit it.

**Lemma 1.2.6.** *Let there be given two spheres $\mathcal{S}(x_1, \lambda_1)$ and $\mathcal{S}(x_2, \lambda_2)$ in a strictly convex normed space $(\mathbb{M}^d, \|\cdot\|)$ with $\lambda_1 > \lambda_2$. Then they touch each other if and only if*

$$\|x_1 - x_2\| = \lambda_1 \pm \lambda_2. \tag{1.6}$$

*Moreover, the sign in (1.6) is plus if the point $x_2$ is an exterior point with respect to $\mathcal{S}(x_1, \lambda_1)$, and it is minus if $x_2$ is an interior point with respect to $\mathcal{S}(x_1, \lambda_1)$.*

The next theorem, which summarizes Proposition 14 and Proposition 41 from [72], gives an information about the number of circles passing through three non-collinear points.

**Theorem 1.2.1.** *The following properties hold in a normed plane $(\mathbb{M}^2, \|\cdot\|)$.*

*(i) If $(\mathbb{M}^2, \|\cdot\|)$ is strictly convex, then through any three non-collinear points at most one circle passes.*

*(ii) If $(\mathbb{M}^2, \|\cdot\|)$ is smooth, then through any three non-collinear points at least one circle passes.*

Since in strictly convex normed planes any planar quadrangle with opposite sides of equal lengths is a parallelogram (see [72, Proposition 12]), the following lemma holds.

**Lemma 1.2.7.** *In a strictly convex normed plane, for any two different circles $\mathcal{C}(y_1, \lambda)$ and $\mathcal{C}(y_2, \lambda)$ with $\mathcal{C}(y_1, \lambda) \cap \mathcal{C}(y_2, \lambda) = \{z_1, z_2\}$ and $z_1 \neq z_2$ the equation $y_1 + y_2 = z_1 + z_2$ holds.*

Let now $K \subset \mathbb{M}^d$ be a *convex body*, i.e., a compact, convex set with nonempty interior. Let the points $p$ and $q$ belong to the boundary of $K$. The segment $[p_1, p_2]$ is called an *affine diameter* of $K$ if there exist two different parallel supporting hyperplanes $H_1$ and $H_2$ of $K$ such that $p_1 \in H_1$ and $p_2 \in H_2$. If all affine diameters of $K$ have the same Minkowskian length, then $K \subset (\mathbb{M}^d, \|\cdot\|)$ is said to be of *constant Minkowskian width*. The *Minkowskian diameter* diam $K$ of a set $K$ is defined by

$$\operatorname{diam} K := \sup\{\|x - y\| : x, y \in K\}.$$

If $K$ is a convex body and $p_1, p_2 \in \operatorname{bd} K$ are such that $\|p_1 - p_2\| = \operatorname{diam} K$, then we call also the segment $[p_1, p_2]$ *diameter* of $K$. Any diameter of $K$ is also an affine diameter, but not vice versa; see [7, Theorem 2, IV].

For each pair $G_1$, $G_2$ of parallel supporting hyperplanes of a convex body $K$ of constant width, every affine diameter generated by $G_1$ and $G_2$ is Birkhoff orthogonal to $G_1$ and $G_2$; see, e.g., [70, Theorem 1, (3)]. This means that for each convex body $K$ of constant width $\lambda$ the distance between every two parallel supporting hyperplanes of $K$ is also $\lambda$.

Let there be given a Jordan curve $\gamma$ in $\mathbb{M}^2$. If the points $p$ and $q$ on $\gamma$ coincide or $p$ precedes $q$ according to a fixed orientation of $\gamma$, we will write $p \prec q$.

At the end, we give three basic definitions from discrete geometry. A collection $\{B_i\}$ of finitely many convex bodies is called a *covering* of the body $B$ if any point of $B$ belongs to $\bigcup_i B_i$ and for every body $B_i$ of $\{B_i\}$ there exists a point $x \in B$ such that $x \in \bigcup_{j \neq i} B_j$. Two convex bodies are *non-overlapping* if they intersect, but have no interior point in common. A family of convex bodies is said to be *packing* if any two members of this family do not overlap each other.

# Chapter 2

# Ball-intersection bodies

Let $(\mathbb{M}^d, \|\cdot\|)$ be a normed space. The intersection of balls of equal radii in $(\mathbb{M}^d, \|\cdot\|)$ is called a *ball-intersection body*. If the number of the balls is finite, then this body is said to be a *ball-polyhedron*. If the points $p_1, \ldots, p_k$ form an *equilateral set* in $(\mathbb{M}^d, \|\cdot\|)$, i.e., $\|p_i - p_j\| = \lambda$ for $i \neq j$ and $i, j \in \{1, \ldots, k\}$, then the ball-polyhedron $\bigcap_{i=1}^{k} \mathcal{B}(p_i, \lambda)$ is said to be an *equilateral ball-polyhedron* with vertices $p_1, \ldots, p_k$.

In this chapter we consider different types of ball-intersection bodies. The fist section is devoted to the so-called ball-hull of a set. It is defined as the intersection of all balls of the same radii containing a given set. We investigate relations between the ball-hull of a set $S$ and the ball-intersection of the same set $S$. The ball-intersection is the intersection of all balls of equal radii centered at $S$, i.e., it is a ball-intersection body, too. In the second section we give a minimal property of a ball-polyhedron that is the intersection of three balls of radius $\lambda$ centered at the vertices of an equilateral triangle of side-length $\lambda$. A subject of the third section is the class of bodies of constant width. As the theorem of Eggleston states, any body of constant width is the ball-intersection of themselves. In this chapter results from the papers [65], [93], and [3] are included. The results from Section 2.1.2 will be submitted. Note that Lemma 2.1.4 and Lemma 2.1.5 are due to Martini and Theorem 2.3.2 and Theorem 2.3.12 to Alonso.

## 2.1 Ball-hull

### 2.1.1 Definition and basic properties

Let there be given a set $K$ in a normed space $(\mathbb{M}^d, \|\cdot\|)$. The *ball-intersection* $\mathcal{BI}(K)$ of $K$ is the intersection of all unit balls whose centers are in $K$, i.e.,

$$\mathcal{BI}(K) := \bigcap_{x \in K} \mathcal{B}(x, 1).$$

The *ball-hull* $\mathcal{BH}(K)$ of $K$ is defined as the intersection of all unit balls that contain $K$. Clearly, $\mathcal{BH}(K) \neq \emptyset$ if and only if $K$ can be covered by a unit ball. Directly from the definitions of the ball-intersection and the ball-hull we get

$$K_1 \subseteq K_2 \implies \mathcal{BI}(K_1) \supseteq \mathcal{BI}(K_2) \text{ and } \mathcal{BH}(K_1) \subseteq \mathcal{BH}(K_2). \quad (2.1)$$

We also note that if $K$ is a set of diameter 1, its ball-hull $\mathcal{BH}(K)$ and its ball-intersection $\mathcal{BI}(K)$ are nonempty sets, and the following implications

$$K \subseteq \mathcal{BH}(K) \subseteq \mathcal{BI}(K) \quad (2.2)$$

hold.

The next propositions give further relations between the ball-intersection and the ball-hull of a set.

**Proposition 2.1.1.** *Let $K$ be a set in a normed space that satisfies $\mathcal{BI}(K) \neq \emptyset$ and $\mathcal{BH}(K) \neq \emptyset$. Then*

*(i) $\mathcal{BH}(K) = \mathcal{BI}(\mathcal{BI}(K))$, and*

*(ii) $\mathcal{BI}(K) = \mathcal{BI}(\mathcal{BH}(K))$.*

*Proof.* (i): $x \in \mathcal{BH}(K) \iff x \in \bigcap_{K \subset \mathcal{B}(p,1)} \mathcal{B}(p,1) \iff$
$x \in \bigcap_{\substack{\|p-q\| \leq 1 \\ \text{for every } q \in K}} \mathcal{B}(p,1) \iff x \in \bigcap_{p \in \mathcal{BI}(K)} \mathcal{B}(p,1) \iff x \in \mathcal{BI}(\mathcal{BI}(K))$.

(ii): $x \in \mathcal{BI}(\mathcal{BH}(K)) \implies x \in \bigcap_{p \in \mathcal{BH}(K)} \mathcal{B}(p,1) \subseteq \bigcap_{p \in K} \mathcal{B}(p,1) = \mathcal{BI}(K)$. For the inverse implication we note that the ball-intersection $\mathcal{BI}(K)$ of a set $K$ can be also viewed as the set of centers of all unit balls that include $K$. Since $K$ and $\mathcal{BH}(K)$ are included in the same unit balls, we obtain (ii). □

**Proposition 2.1.2.** *If* diam $K =$ diam $\mathcal{BI}(K) = 1$, *then the ball-intersection and the ball-hull of $K$ coincide.*

*Proof.* If diam $\mathcal{BI}(K) = 1$, then by the implication (2.2) and (i) we get $\mathcal{BI}(K) \subseteq \mathcal{BI}(\mathcal{BI}(K)) = \mathcal{BH}(K)$, yielding $\mathcal{BI}(K) = \mathcal{BH}(K)$, again by (2.2). □

Now we give an example for the ball-hull of the set $\{p_1, p_2, p_3\}$, where the points $p_1, p_2, p_3$ form an equilateral triangle in a normed plane $(\mathbb{M}^2, \|\cdot\|)$. The ball-intersection $\mathcal{BI}(\{p_1, p_2, p_3\})$ (see Figure 2.1) is, in fact, a *Minkowskian Reuleaux triangle*. Note that in normed planes a

## 2.1. BALL-HULL

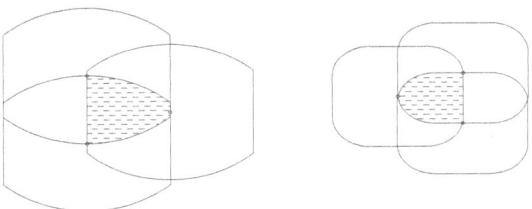

Figure 2.1: Reuleaux triangles in normed planes.

Reuleaux triangle $K$ was defined in [87] and [88] as the figure bounded by three arcs of radius $\lambda \in \mathbb{R}^+$ which are centered at the vertices of an equilateral triangle with sides of length $\lambda$, i.e.,

$$K = \left( \bigcup_{\substack{i,j \in \{1,2,3\} \\ \text{and } i \neq j}} \operatorname{conv} \operatorname{arc}\{p_i, p_j; p_k\} \right) \bigcup \operatorname{conv} \{p_1, p_2, p_3\}. \tag{2.3}$$

But one can see that this is the ball-intersection of $\{p_1, p_2, p_3\}$; for a detailed proof of this fact we refer to [29, p. 18, Proposition 1.6]. To clarify how the ball-hull of $\{p_1, p_2, p_3\}$ looks like, the following lemma is necessary. This lemma will also be used in the considerations of the next sections.

**Lemma 2.1.1.** *Let $(\mathbb{M}^2, \|\cdot\|)$ be a normed plane with unit disc $\mathcal{B}$. If the points $p$ and $q$ belong to $\mathcal{B}$ such that $\|p - q\| = 1$, then any circular arc of radius 1 and with endpoints $p$ and $q$ also belongs to $\mathcal{B}$.*

*Proof.* The possible centers of circular arcs of radius 1 with endpoints $p$ and $q$ are exactly two (in different half planes with respect to the line $\mathrm{L}(p,q)$) if and only if the longest segment in the unit circle parallel to $\mathrm{L}(p,q)$ has length at most 1; see, e.g., [72, Proposition 33]. But if the length of this longest segment is more than 1, then all arcs of radius 1 joining $p$ and $q$ coincide with the line segment $[p,q]$. Thus, in this case our statement holds. Let us now consider the case when there exist exactly two points $x_1$ and $x_2$ such that the circles $\mathcal{C}(x_1, 1)$ and $\mathcal{C}(x_2, 1)$ pass through $p$ and $q$. Let the line $\mathrm{L}(p,q)$ not pass through the origin 0. Let $x_1$ lie in this half plane bounded by $\mathrm{L}(p,q)$ that contains 0. Assume also that both the points $p$ and $q$ do not lie on the unit circle $\mathcal{C}$, say $p \notin \mathcal{C}$, i.e. $\|p\| < 1$. For $x_1$ there exist the following possibilities:

1. $x_1 \in \operatorname{conv} \{p, q, 0\}$;

2. $x_1 \in \mathrm{HS}_0^+(p,q) \cap \mathrm{HS}_q^-(p,0) \cap \mathrm{HS}_p^+(0,q)$ or $x_1 \in \mathrm{HS}_0^+(p,q) \cap \mathrm{HS}_p^-(q,0) \cap \mathrm{HS}_q^+(0,p)$;

3. $x_1 \in \mathrm{HS}_q^-(p,0) \cap \mathrm{HS}_p^-(0,q)$.

Case 1. is impossible. Otherwise $2 = \|p - x_1\| + \|x_1 - q\| \leq \|p\| + \|q\| < 2$, a contradiction to Lemma 1.2.2. If the first case of 2. holds, assume that there exists a point $x$ on the circular arc $\mathrm{arc}(p,q;x_1)$ that does not belong to $\mathcal{C}$, i.e., $\|x\| > 1$. Since $x$ belongs to $\mathrm{HS}_q^+(x_1,p) \cap \mathrm{HS}_p^+(x_1,q)$, the points $p, x_1, 0, x$ form a convex quadrangle. Thus, by Lemma 1.2.4 we have $2 \geq \|x - x_1\| + \|p\| \geq \|p - x_1\| + \|x\| > 2$, a contradiction. For the second case of 2. we consider the convex quadrangle with vertices $q, x, 0, x_1$, and again we get a contradiction. For case 3. we denote by $y$ the intersection point of the ray $\mathrm{R}_0^+(x_1)$ and the circular arc $\mathrm{arc}(p,q;x_1)$. Then $\|y\| \leq 1$. Now we consider a point $x$ from $\mathrm{arc}(p,q;x_1)$ different to $y$. Thus by Lemma 1.2.3 we have $\|x\| \leq \max\{\|x - y\|, \|x - x_1\| = 1\}$. Since the monotonicity lemma yields $\|x - y\| \leq \|p - q\| = 1$, we get $\|x\| \leq 1$. Moreover, from the fact that the only possible situation is 3., we conclude the implication $0 \in \mathrm{conv}\{p, q, x_1\}$. On the other hand, the uniqueness of the centers $x_1$ and $x_2$ implies that the points $p, x_1, q, x_2$ form a parallelogram. This means that $0$ is contained in the cone $\mathrm{HS}_q^+(x_2, p) \cap \mathrm{HS}_p^+(x_2, q)$. Let $x$ be an arbitrary point from $\mathrm{arc}(p,q;x_2)$. We distinguish the following cases:

(i) $x \in \mathrm{conv}\{p, q, 0\}$;

(ii) $x \in \mathrm{HS}_0^+(p,q) \cap \mathrm{HS}_q^-(p,0) \cap \mathrm{HS}_p^+(0,q)$ or $x_1 \in \mathrm{HS}_0^+(p,q) \cap \mathrm{HS}_p^-(q,0) \cap \mathrm{HS}_q^+(0,p)$;

(iii) $x \in \mathrm{HS}_q^-(p,0) \cap \mathrm{HS}_p^-(0,q)$.

It is our aim to prove that in all three cases $\|x\| \leq 1$. If (i) holds and $\mathrm{R}_x^+(0) \cap [p,q] = \{y\}$, then Lemma 1.2.3 yields $\|x\| \leq \|y\| \leq \max\{\|p\|, \|q\|\} \leq 1$. For the first case of (ii) we consider the points $p, x, 0, q$ that form a convex quadrangle. If $\|x\| > 1$, then $2 < \|x\| + \|p - q\| \leq \|p\| + \|x - q\| \leq 2$, a contradiction to Lemma 1.2.4. For (iii) denote by $z$ the intersection point of $\mathrm{R}_0^+(x_2)$ and $\mathrm{arc}(p, q; x_2)$. If $x = y$, then $\|x\| \leq 1$. Otherwise, we have $\|x\| \leq \max\{\|x - z\|, \|x - x_2\|\} = 1$. Thus it remains to consider the case when $p$ and $q$ lie on $\mathcal{C}$. Then one of the two centers, say $x_1$, has to coincide with the origin $0$. In this case our statement follows from Lemma 1.2.5, which says that the part of $\mathcal{C}(x_2, 1)$ lying in the half plane $\mathrm{HS}_0^+(p,q)$, in fact the arc $\mathrm{arc}(p,q;x_2)$, belongs to the convex hull of the part of $\mathcal{C}$ that lies in $\mathrm{HS}_0^+(p,q)$. □

**Remark 2.1.1.** *The next example shows that Lemma 2.1.1 is no longer true in dimensions $\geq 3$. Let $(\mathbb{M}^3, \|\cdot\|)$ be the normed space whose unit ball $\mathcal{B}$ is the regular octahedron with vertices $(\pm 1, 0, 0), (0, \pm 1, 0), (0, 0, \pm 1)$. Then the circular arc of radius $1$ with center $(0, \frac{1}{2}, \frac{1}{2})$ and endpoints $(\frac{1}{2}, 0, \frac{1}{2}), (-\frac{1}{2}, 0, \frac{1}{2})$ does not belong to $\mathcal{B}$.*

**Proposition 2.1.3.** *Let $p_1, p_2, p_3$ be three points in a normed plane with $\|p_i - p_j\| = 1$ for $i, j \in \{1, 2, 3\}$, $i \neq j$. The ball-hull $\mathcal{BH}(\{p_1, p_2, p_3\})$ is the Reuleaux triangle with vertices $p_1, p_2$, and $p_3$.*

## 2.1. Ball-hull

*Proof.* Let $\mathcal{B}(p,1)$ be an arbitrary disc containing the points $p_1, p_2$, and $p_3$. Then, according to Lemma 2.1.1 and (2.3), it contains $\cap_{i=1}^{3}\mathcal{B}(p_i,1)$. Since diam $\{p_1, p_2, p_3\} = 1$ the implication (2.2) yields

$$\mathcal{BH}(\{p_1, p_2, p_3\}) = \bigcap_{i=1}^{3}\mathcal{B}(p_i, 1).$$

□

The next notion, namely that of the *ball-intersection property*, is well known in Euclidean spaces as well as in normed spaces. Via the ball-intersection property an important description of the class of bodies of constant width in Euclidean spaces and of the class of complete bodies in normed spaces is possible. A convex body $K$ of diameter 1 has the *ball-intersection property* if $\mathcal{BI}(K) = K$. The following classical theorems clarify the relations between the ball-intersection property, constant width, and completeness; see [57], [36], and the surveys [32], [70, pp. 98-99]. Note that in a normed space $(\mathbb{M}^d, \|\cdot\|)$ a convex body $K$ is called *complete* if diam $(K \cup \{x\}) >$ diam $K$ for every $x \notin K$.

**Theorem 2.1.1.** [Meissner[1]] *Let $K$ be a body of diameter 1 in a Euclidean space. Then the following statements are equivalent.*

1. *$K$ is of constant width.*

2. *$K$ is complete.*

3. *$K$ has the ball-intersection property.*

**Theorem 2.1.2.** [Kelly, [57]] *A convex body $K$ of diameter 1 in a normed plane has the ball-intersection property if and only if it is of constant Minkowskian width.*

**Theorem 2.1.3.** [Eggleston, [36]] *Let $K$ be a body of diameter 1 in a normed space $(\mathbb{M}^d, \|\cdot\|)$.*

1. *If $K$ is of constant width, then it has the ball-intersection property.*

2. *$K$ is complete if and only if it has the ball intersection property.*

Now we say that a convex body $K$ of diameter 1 has the *ball-hull property* if $\mathcal{BH}(K) = K$. Implication (2.2) implies that the class of bodies having the ball-hull property is wider than the class of bodies with ball-intersection property. The following example shows that the both classes do not coincide.

---
[1] In fact, usually the equivalence of 1. and 2. if referred to as the theorem of Meissner.

**Example 2.1.1.** In a normed plane $(\mathbb{M}^2, \|\cdot\|)$, let there be given two points $p$ and $q$ with $\|p - q\| = 1$. Let $\mathcal{C}(x, 1)$ be a circle passing through $p$ and $q$. If $y = p + q - x$, then the circle $\mathcal{C}(y, 1)$ also passes though $p$ and $q$. Let $K = \mathcal{D}(x, 1) \cap \mathcal{D}(y, 1)$. Due to Lemma 2.1.1 the ball-hull of $K$ is $\mathcal{D}(x, 1) \cap \mathcal{D}(y, 1)$, i.e., $K$ has the ball-hull property. But since $K$ is centrally symmetric, it is not of constant width, i.e., it does not posses the ball-intersection property.

Again returning to implication (2.2), we see that any convex body that has the ball-intersection property also has the hull-intersection property, i.e., the following theorem holds.

**Theorem 2.1.4.** *In a normed space $(\mathbb{M}^d, \|\cdot\|)$ any complete body of diameter 1 has the ball-hull property.*

Setting a condition additional to completeness, we can prove that the converse of Theorem 2.1.4 holds, too.

**Theorem 2.1.5.** *If a convex body $K$ of diameter 1 in a normed space $(\mathbb{M}^d, \|\cdot\|)$ has the ball-hull property and $\operatorname{diam} \mathcal{BI}(K) = 1$ is satisfied, then $K$ is complete.*

*Proof.* For any convex body $P$ of diameter 1 we have $P \subseteq \mathcal{BI}(P)$. Applying this for $P = \mathcal{BI}(K)$, we get $\mathcal{BI}(K) \subseteq \mathcal{BI}(\mathcal{BI}(K))$. On the other hand, Proposition 2.1.1 implies $\mathcal{BI}(\mathcal{BI}(K)) = \mathcal{BH}(K)$. Therefore $\mathcal{BI}(K) \subseteq \mathcal{BH}(K) = K$. Since $\operatorname{diam} K = 1$, also the relation $K \subseteq \mathcal{BI}(K)$ holds. Thus we get the desired statement. □

### 2.1.2 Meissner's bodies

As we have seen in the previous section, Reuleaux triangles are classical examples of bodies of constant width being ball-polyhedra. But in [50] it is proved that in a Euclidean space of dimension $\geq 3$ no finite intersection of balls has constant width, unless it reduces to a single ball, i.e., in $\mathbb{E}^d$, $d \geq 3$, there exist no trivial ball-polyhedra of constant width. It seems that the *Reuleaux tetrahedron* defined as the ball-intersection of $\{p_1, \ldots, p_4\}$, where $\|p_i - p_j\| = \lambda$ for $i, j \in \{1, \ldots, 4\}$ and $i \neq j$, can play this role. But this body is not of constant width; see [73]. Meissner was able to construct a body, usually called *Meissner tetrahedron*, which is of constant width and very close to the Reuleaux tetrahedron. Meissner's construction is obtained by flipping couples of edges of the Reuleaux tetrahedra. In [50] Lachand-Robert and Oudet gave an inductiv construction of bodies of constant width for Euclidean space of arbitrary dimension. A variant of this construction yields, in the two-dimensional case, a Reuleuax triangle, and in the three-dimensional case a Meissner tetrahedron is obtained. For that reason they called them *Meissner's bodies*. In this section we apply this construction to normed spaces and prove that the resulting body is complete. We prove also that, although Meissner's bodies in $\mathbb{E}^d$ are no ball-polyhedra, they can be approximated by ball-polyhedra, namely by the ball-hull and the

## 2.1. BALL-HULL

ball-intersection of a finite point set that has a special position with respect to the considered Meissner body.

We start with the construction of Lachand-Robert and Oudet. Let HP be a hyperplane in $\mathbb{E}^d$, let HS$^+$ and HS$^-$ be the two open half-spaces bounded by HP, and let $K_0 \subset$ HP be a $(d-1)$-dimensional convex body of constant width $\lambda$. Let $Q$ be any set satisfying

$$K_0 \subseteq Q \subset \text{cl HS}^- \cap \left( \bigcap_{x \in K_0} \mathcal{B}(x, \lambda) \right).$$

Then the body $K = K^+ \cup K^-$, where

$$K^+ := \text{cl HS}^+ \cap \left( \bigcap_{x \in Q} \mathcal{B}(x, \lambda) \right) \text{ and } K^- := \text{HS}^- \cap \left( \bigcap_{x \in K^+} \mathcal{B}(x, \lambda) \right), \quad (2.4)$$

is of constant width. Let now $(\mathbb{M}^d, \|\cdot\|)$ be a normed space. The resulting body of the same construction for $Q = K_0$ we call a *Meissner body induced by $K_0$*.

**Theorem 2.1.6.** *In a normed space $(\mathbb{M}^d, \|\cdot\|)$, let $K_0$ be an $(d-1)$-dimensional complete body lying in the hyperplane* HP. *Then the Meissner body $K$ induced by $K_0$ is complete.*

*Proof.* Without loss of generality we assume that diam $K_0 = 1$. First we prove that also the diameter of $K$ is 1. Let $p, q$ be two distinct points of $K$. We distinguish the following cases:

1. $p, q \in K_0$;
2. $p \in K_0$, $q \in K^+ \setminus K_0$;
3. $p, q \in K^+ \setminus K_0$;
4. $p \in K^+$, $q \in K^-$;
5. $p, q \in K^-$.

In view of (2.4) we have directly in 1., 2., and 4. that $\|p - q\| \leq 1$. For 3. we consider an affine diameter $[p_1, q_1]$ of $K_0$ such that $p, q, p_1, q_1$ lie in a 2-flat $G$. Since the ball $\mathcal{B}(p_1, 1)$ contains $p$ and $q$, the disc $\mathcal{D}(p_1, 1)$ which is the intersection of $\mathcal{B}(p_1, 1)$ and $G$ also contains $p$ and $q$. Moreover, $p$ and $q$ belong to the same half-plane bounded by the line $\text{L}(p_1, q_1)$. We may assume that $p$ and $q$ lie on the boundary of this disc. We can also rename the points $p$ and $q$ such that $p, q, q_1, p_1$ form a convex quadrangle in this order. Then by Lemma 1.2.4 we get

$$2 \geq \|p_1 - q\| + \|p - q_1\| \geq \|p_1 - q_1\| + \|p - q\| = 1 + \|p - q\|.$$

The same considerations show that in 5. the distance between $p$ and $q$ is also $\leq 1$. Thus by (2.2) we get $K \subset \mathcal{BI}(K)$. To prove the converse implication we consider an arbitrary point $p$ from $\mathcal{BI}(K)$. Then the claims

(i) $\|p - x\| \leq 1$ for all $x \in K_0$ and

(ii) $\|p - y\| \leq 1$ for all $y \in K^+$

hold. If $p \in \mathrm{HP}$, then (i) implies that $p$ belongs to all $(d-2)$-balls centered at $K_0$. Since $K_0$ is complete, the point $p$ is contained in $K_0 \subset K$. If $p$ lies in $\mathrm{HS}^+$, then by (i) we have $p \in \mathcal{BI}(K_0)$, yielding $p \in K^+ \subset K$. In the case $p \in \mathrm{HS}^-$, by (ii) we also obtain $x \in K^- \subset K$. Therefore $\mathcal{BI}(K) = K$, which is sufficient to the claim that $K$ is complete. □

**Remark 2.1.2.** *Theorem 2.1.6 is a generalization of Theorem 4.1 in [50] (for the case $Q = K_0$) which says the same, but for the Euclidean subcase. If we restrict our Theorem 2.1.6 to Euclidean space, we have a new proof of Theorem 4.1 in [50].*

Now we consider a Meissner body in a Euclidean space $\mathbb{E}^3$ starting from a Reuleaux triangle $K_0$ with vertices $p_1, p_2, p_3$ in a 2-plane. Assume that diam $K_0 = 1$. Induced by this Reuleaux triangle we obtain the Meissner tetrahedron $K$. Let $p_4$ be a point lying in that half space with respect to $\mathrm{HP}(p_1, p_2, p_3)$ which contains $K^+$ such that $\|p_4 - p_i\| = 1$. We call the equilateral ball-polyhedron with vertices $p_1, p_2, p_3, p_4$ *associated with* $K$. We shall prove that any Meissner tetrahedron, which is created from a Reuleaux triangle, can be approximated by its associated equilateral ball-polyhedron and the ball-hull of the vertices of this equilateral ball-polyhedron. For this goal we need the following

**Lemma 2.1.2.** *Let $K$ be the Meissner tetrahedron in $\mathbb{E}^3$ induced by the Reuleaux triangle $K_0$. If $p_1, p_2, p_3, p_4$ are the vertices of the equilateral ball-polyhedron associated with $K$, then $p_4 \in K$.*

*Proof.* Let diam $K = 1$. If we prove that for any $x \in K_0$ the distance from $p_4$ to $x$ is at most 1, then the lemma is true. If $x$ belongs to the boundary of conv $\{p_1, p_2, p_3\}$, this is trivial. If $x \in \mathrm{int\ conv\ } \{p_1, p_2, p_3\}$, we consider $x'$ determined as the intersection of the ray $\mathrm{R}_x^+(p_3)$ and the segment $[p_1, p_2]$. Thus by Lemma 1.2.3 we get

$$\|p_4 - x\| < \max\{\|p_4 - p_3\|, \|p_4 - x'\|\} = 1.$$

So it remains to consider the case when $x \in K_0 \setminus \mathrm{conv\ } \{p_1, p_2, p_3\}$. In view of the above proof it is sufficient to choose $x$ on the boundary of $K_0$, say $x \in \mathrm{arc}(p_1, p_2; p_3)$. Since any arc of radius 1 belongs to the ball $\mathcal{B}(p_4, 1)$ if its endpoints are contained in $\mathcal{B}(p_4, 1)$ (see, e.g., [100, p. 373, Theorem 7.6.4]), we again obtain that $\|p_4 - x\| \leq 1$.

□

**Theorem 2.1.7.** *If $K$ is the Meissner tetrahedron in $\mathbb{E}^3$ induced by a Reuleaux triangle $K_0$ and the associated ball-polyhedron of $K$ has the vertices $p_1, p_2, p_3, p_4$, then*

$$\mathcal{BH}(p_1, \ldots, p_4) \subset K \subset \mathcal{BI}(p_1, \ldots, p_4). \tag{2.5}$$

2.1. BALL-HULL

*Proof.* Since
$$K \subset \mathcal{BI}(p_1, \ldots, p_4), \tag{2.6}$$
it follows by (2.1) and Proposition 2.1.1 that
$$\mathcal{BI}(K) \supseteq \mathcal{BI}(\mathcal{BI}(p_1, \ldots, p_4)) = \mathcal{BH}(p_1, \ldots, p_4). \tag{2.7}$$
On the other hand, $K$ is of constant width, i.e., $K = \mathcal{BI}(K)$. Thus by (2.6) and (2.7) we get the implications (2.5). □

**Remark 2.1.3.** *Due to the proof of Lemma 2.1.2 it also holds in a normed space with unit ball $\mathcal{B}$ having the following property:*

> for any two points $p, q$ belonging to $\mathcal{B}$ with $\|p - q\| = 1$ all circular arcs of radius 1 with endpoints $p$ and $q$ also belong to $\mathcal{B}$.

*Thus Theorem 2.1.7 is also true in three-dimensional normed spaces having the above property. Note that due to our Remark 2.1.1 not all normed spaces of dimension 3 have this property.*

### 2.1.3 Relations between the ball-hull and the ball-intersection of a convex body

Now we prove one more relation between the ball-hull and the ball-intersection of a set $K$ in strictly convex normed planes. Namely, the Minkowski sum of the ball-hull and the ball-intersection of a set of diameter 1 is a convex set constant Minkowskian width 2. The Euclidean version of this statement was given by Capoyleas [28] and Bezdek, Connely, Csikós [16].

First we establish a few lemmas.

**Lemma 2.1.3.** *In a strictly convex normed plane $(\mathbb{M}^2, \|\cdot\|)$, let there be given a rhombus $\mathcal{Q}(p_1, p_2, p_3, p_4)$ with side-lengths 1. If $x \in \mathcal{D}(p_1, 1) \cap \mathcal{D}(p_3, 1)$ and $y \in \mathcal{D}(p_2, 1) \cap \mathcal{D}(p_4, 1)$, then $\|x - y\| \leq 1$.*

*Proof.* We distinguish two cases.

I. $\mathcal{D}(p_1, 1) \cap \mathcal{D}(p_3, 1) \subset \mathcal{D}(p_2, 1) \cap \mathcal{D}(p_4, 1)$; see Figure 2.2. Then we have $p_2 \in \mathcal{C}(p_1, 1) \cap \mathcal{C}(p_3, 1) \subset \mathcal{D}(p_1, 1) \cap \mathcal{D}(p_3, 1) \subset \mathcal{D}(p_4, 1)$. Therefore

$$\|p_2 - p_4\| \leq 1. \tag{2.8}$$

If the diagonals of $\mathcal{Q}(p_1, p_2, p_3, p_4)$ intersect in $z$, then the triangle inequality, applied to the triangle $\mathcal{T}(p_1, z, p_2)$, yields $\frac{1}{2}\|p_1 - p_3\| + \frac{1}{2}\|p_2 - p_4\| > 1$. Therefore

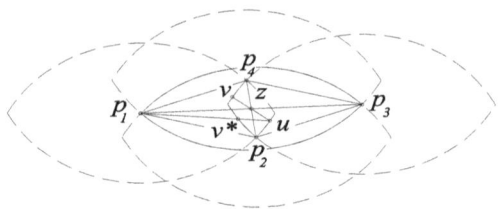

Figure 2.2: The proof of Lemma 2.1.3, I.

$$\|p_1 - p_3\| > 1, \tag{2.9}$$

i.e., $p_1$ is an exterior point of $D = \mathcal{D}(p_1, 1) \cap \mathcal{D}(p_3, 1)$. Assume that $x, y \in D$, and $[u, v]$ be a diameter of $D$. Due to the monotonicity lemma the points $u$ and $v$ do not lie in the same half-plane with respect to the line $\mathrm{L}(p_2, p_4)$. Moreover, since $D$ is symmetric with respect to the midpoint $z$ of $[p_2, p_4]$, the diameter $[u, v]$ passes through $z$. Let $u \in \mathcal{C}(p_1, 1)$, $v \in \mathcal{C}(p_3, 1)$. The case that $u$, $v$, and $p_1$ are collinear yields $\|u - v\| < 1$. If $p_1 \notin \mathrm{L}(u, v)$, assume that $[u, v] \neq [p_2, p_4]$. If $[p_1, u] \cap \mathcal{C}(p_3, 1) = \{v^*\}$, then $u$, $p_2$, $v^*$, $v$ (or $u$, $p_4$, $v^*$, $v$) form a convex quadrangle. This means that the quadrangle $\mathcal{Q}(u, p_2, p_1, v)$ (or $\mathcal{Q}(u, p_4, p_1, v)$) is convex, too. Thus by Lemma 1.2.4, we get

$$\|p_2 - v\| + \|u - p_1\| > \|p_1 - p_2\| + \|u - v\| \iff \|p_2 - v\| > \|u - v\|,$$

a contradiction to the fact that $[u, v]$ is a diameter. Therefore $[u, v] \equiv [p_2, p_4]$. This means that diam $D \leq 1$ (see (2.9)) and $\|x - y\| \leq 1$.

Now we consider the case that $y \notin \mathcal{D}(p_1, 1) \cap \mathcal{D}(p_3, 1)$. Denote $\mathcal{D}(p_2, 1) \cap \mathcal{D}(p_4, 1)$ by $D'$. Let $D_1 = (\mathrm{HS}^+_{p_2}(p_1, x) \cap \mathrm{HS}^+_{p_1}(p_2, x)) \cap D'$, $D_2 = (\mathrm{HS}^+_{p_2}(p_3, x) \cap \mathrm{HS}^+_{p_3}(p_2, x)) \cap D'$, $D_3 = (\mathrm{HS}^+_{p_3}(p_4, x) \cap \mathrm{HS}^+_{p_4}(p_3, x)) \cap D'$, $D_4 = (\mathrm{HS}^+_{p_4}(p_1, x) \cap \mathrm{HS}^+_{p_1}(p_4, x)) \cap D'$, and $y \in D_1$, say. Then the points $p_1, y, x, p_4$ (or the points $p_3, y, x, p_4$ when $y \notin (\mathrm{HS}^+_{p_1}(p_4, x) \cap \mathrm{HS}^+_{p_4}(p_1, x)) \cap D_1$) form a convex quadrangle, and

$$\|x - y\| + \|p_1 - p_4\| < \|p_4 - y\| + \|p_1 - x\|$$

holds, by Lemma 1.2.4. Thus we get $\|x - y\| + 1 < 2 \iff \|x - y\| < 1$.

II. $\mathcal{D}(p_1, 1) \cap \mathcal{D}(p_3, 1) \not\subset \mathcal{D}(p_2, 1) \cap \mathcal{D}(p_4, 1)$; see Figure 2.3. Let now $x$ and $y$ be placed as in Figure 2.3 or $x, y \in \bigcap_{i=1}^{4} \mathcal{D}(p_i, 1)$. Denote by $t$ the intersection point of $\mathrm{arc}(p_1, p_3; p_2)$ and

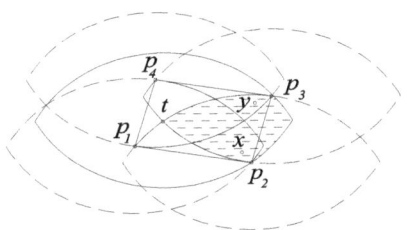

Figure 2.3: The proof of Lemma 2.1.3, II.

arc$(p_2, p_4; p_3)$. Then $D(t,1) \cap D(p_2,1) \cap D(p_3,1)$ is a Reuleaux triangle. If $y \in \text{conv}\{p_4, t, p_3\}$, then the monotonicity lemma implies that $\|t - y\| < 1$. If $y \notin \text{conv}\{p_4, t, p_3\}$, then the points $p_4, t, y, p_3$ form a convex quadrangle. Therefore

$$\|p_4 - y\| + \|t - p_3\| > \|t - y\| + \|p_3 - p_4\| \iff \|p_4 - y\| > \|t - y\|,$$

again by Lemma 1.2.4. Since $\|p_4 - y\| \le 1$, it follows that $\|t - y\| < 1$. Analogously, $\|t - x\| < 1$. Thus we get that the Reuleaux triangle $D(t,1) \cap D(p_2,1) \cap D(p_3,1)$ contains the points $x$ and $y$. Hence $\|x - y\| \le 1$. □

**Lemma 2.1.4.** *In a strictly convex normed plane, let there be given two points $p, q$ with $\|p - q\| \le 2$. If $C(x_1, 1)$ and $C(x_2, 1)$ are the circles passing through $p$ and $q$, then we have $\mathcal{BH}([p, q]) = D(x_1, 1) \cap D(x_2, 1)$.*

*Proof.* Clearly, $\mathcal{BH}([p, q]) \subseteq D(x_1, 1) \cap D(x_2, 1)$. Let $D(y, 1) \supset [p, q]$. Then

$$p, q \in D(y, 1) \iff y \in D(p, 1) \cap D(q, 1).$$

Let $z \in D(x_1, 1) \cap D(x_2, 1)$. We have that the points $p, x_1, q, x_2$ form a rhombus of side-lengths 1. (Note that in a strictly convex Minkowski plane a quadrangle with opposite sides of equal length is a parallelogram; see [72, Proposition 12].) Then Lemma 2.1.4 implies $\|z - y\| \le 1$, i.e., $D(x_1, 1) \cap D(x_2, 1) \subseteq D(y, 1)$. Thus we get $\mathcal{BH}([p, q]) = D(x_1, 1) \cap D(x_2, 1)$. □

**Lemma 2.1.5.** *In a strictly convex plane $(\mathbb{M}^2, \|\cdot\|)$, let the points $p$ and $q$ belong to the unit disc $\mathcal{D}$. If $C(x, 1)$ is a circle passing through $p$ and $q$, then the arc* arc$(p, q; x)$ *also belong to $\mathcal{D}$.*

*Proof.* Let $\varphi$ be the translation mapping $C(x, 1)$ into $C = \text{bd}\,\mathcal{D}$. If $C(x,1) \cap C = \{p_1, p_2\}$ and $\gamma_2$ is the part of $C(x, 1)$ on the same side as $\varphi(p_1)$, then according to Lemma 1.2.5 $\gamma_2 \subset \mathcal{D}$. Since the arc arc$(p, q; x)$ is a part of $\gamma_2$, we get arc$(p, q; x) \subset \mathcal{D}$. □

**Lemma 2.1.6.** *In a strictly convex normed plane, let $K$ be a set with $\mathcal{BH}(K) \neq \emptyset$. If $p, q \in \mathcal{BH}(K)$, then $\mathcal{BH}([p, q]) \subseteq \mathcal{BH}(K)$.*

*Proof.* Let $\mathcal{B}(x, 1)$ be a disc containing $K$. Then $\mathcal{B}(x, 1)$ contains $p$ and $q$ and by Lemma 2.1.4 and Lemma 2.1.5 it contains $\mathcal{BH}([p, q])$, too. □

Now we are ready to establish the main result in this subsection.

**Theorem 2.1.8.** *In a strictly convex normed plane, let there be given a set $K$ of diameter 1. Then the Minkowski sum $\mathcal{BI}(K) + \mathcal{BH}(K)$ is a convex set of constant Minkowskian width 2.*

*Proof.* For an arbitrary direction, denote by $G_1, G_2$ the supporting lines of $\mathcal{BI}(K)$ parallel to this direction. Let $H_1, H_2$ be the supporting lines of $\mathcal{BH}(K)$ also parallel to $G_1$. According to the implication (2.2) we have that the lines $H_1, H_2$ belong to the strip with bounding lines $G_1, G_2$. Let them be placed in the order $G_1, H_1, H_2, G_2$. Let $x_i \in G_i \cap \mathcal{BI}(K), y_i \in H_i \cap \mathcal{BH}(K)$, $i = 1, 2$. Since $x_2 \in \mathcal{BI}(K)$, we get

$$\|x_2 - p\| \leq 1 \text{ for any } p \in K \iff K \subset \mathcal{D}(x_2, 1).$$

But $y_1 \in \mathcal{BH}(K)$, which means that $y_1 \in \mathcal{D}(x_2, 1) \iff \|y_1 - x_2\| \leq 1$. Thus we obtain

$$d(H_1, G_2) \leq \|y_1 - x_2\| \leq 1. \tag{2.10}$$

Consider a circle $\mathcal{C}(x, 1)$ through $y_1$ with $H_1$ as supporting line at $y_1$. Let $\mathcal{C}(x, 1)$ also lie in $\operatorname{HS}_{G_2}^+(H_1)$, i.e., in the half-plane bounded by $H_1$ and containing $G_2$. Choose a point $y \in \mathcal{BH}(K)$. Since $y_1$ also belongs to $\mathcal{BH}(K)$, the monotonicity lemma implies that $\|y - y_1\| \leq 2$. If $y \in \operatorname{L}(x, y_1)$, then $\|x - y\| \leq 1$, i.e., $y \in \mathcal{D}(x, 1)$. Let now $y \notin \operatorname{L}(x, y_1)$. Assume that $\|x - y\| > 1$. We have that $\mathcal{BH}([y_1, y]) \neq \emptyset$. Let $\mathcal{C}(x^*, 1)$ and $\mathcal{C}(x^{**}, 1)$ denote the circles of radius 1 passing through $y_1$ and $y$, i.e., $\mathcal{BH}([y_1, y]) = \mathcal{D}(x^*, 1) \cap \mathcal{D}(x^{**}, 1)$. Clearly, $x^*$ and $x^{**}$ lie in different half-planes with respect to $\operatorname{L}(y_1, y)$. The point $x^*$ is contained in the double cone of $y_1$ and $y$ with apex $x^{**}$ (see [72, Proposition 17]). If $x^{**}$ belongs to the half-plane opposite to $\operatorname{HS}_{G_2}^+(H_1)$, then $x^{**}$ and $y$ lie in different half-planes with respect $H_1$, and $x^*$ has to lie in $\operatorname{HS}_{G_2}^+(H_1)$. Therefore at least one of the points $x^*$ and $x^{**}$ belongs to $\operatorname{HS}_{G_2}^+(H_1)$. Let $x^* \in \operatorname{HS}_{G_2}^+(H_1) \cap \operatorname{HS}_x^+(y_1, y)$, say. Since $\|y - x\| + \|x - y_1\| > 2 = \|x^* - y\| + \|x^* - y_1\|$, the point $x$ does not belong to conv $\{y, x^*, y_1\}$; see Lemma 1.2.2. Assume that $x \in \operatorname{HS}_y^+(y_1, x^*)$. Then the quadrangle with vertices $x, x^*, y_1$, and $y$ is convex and $2 = \|x - y_1\| + \|y - x^*\|$, $\|x^* - y_1\| + \|x - y\| > 2$, a contradiction to Lemma 1.2.4. Therefore $x \in \operatorname{HS}_y^-(y_1, x^*) \cap \operatorname{HS}_{G_2}^+(H_1) \cap \operatorname{HS}_x^+(y_1, y)$. Since $\|x - y\| > 1$, the point $x^*$ does not coincide with $x$. This means that the line $H_1$ is not a supporting line of $\mathcal{C}(x^*, 1)$. Hence there exists a point $z \in \mathcal{C}(x^*, 1) \cap H_1$ different to $y_1$. Assume that $z \in \operatorname{HS}_y^+(y_1, x^*)$. In view of $\|x - y_1\| + \|x - z\| > 2 = \|x^* - y_1\| + \|x^* - z\|$ we get

$x \notin \operatorname{conv} \{x^*, z, y_1\}$ (see again Lemma 1.2.2), i.e., the quadrangle with vertices $x, z, y_1$, and $x^*$ is convex. Therefore

$$2 = \|x - y_1\| + \|x^* - z\| > \|x - z\| + \|x^* - y_1\| > 2,$$

a contradiction. Thus we have proved that a part of $\mathcal{D}(x^*, 1) \cap \mathcal{D}(x^{**}, 1)$ belongs to $\operatorname{HS}^-_{G_2}(H_1)$. Since $y_1, y \in \mathcal{BH}(K)$, this contradicts Lemma 2.1.6. So we get that $y \in \mathcal{D}(x, 1)$, and therefore $\mathcal{D}(x, 1)$ covers $\mathcal{BH}(K)$. But $K \subseteq \mathcal{BH}(K)$, which means that $x \in \mathcal{BI}(K)$ and that $x$ lies in the strip with bounding lines $H_1$ and $G_2$. Let $\operatorname{L}(y_1, x) \cap G_2 = \{x'\}$. Then

$$d(H_1, G_2) = \|y_1 - x'\| \geq \|y_1 - x\| = 1. \tag{2.11}$$

By the inequalities (2.10) and (2.11) we obtain that $d(H_1, G_2) = 1$. In the same way we get that also $d(G_1, H_2) = 1$. Since the Minkowskian width of the sum of two sets in a given direction is the sum of the Minkowskian widths of these two sets in the same direction, the proof is complete. $\square$

## 2.2   On a theorem of Chakerian

If $\mathcal{RT}$ is a Reuleaux triangle in the Euclidean plane of width $\lambda$ and any congruent copy $P'$ of a compact, convex set $P$ can be covered by a translate of $\mathcal{RT}$, then $P$ can also be covered by a translate of an arbitrary convex body of constant width $\lambda$. This statement is known as *Chakerian's theorem* and was proved in [31]. Another proof was given later by Bezdek and Connelly; see [14]. In this section we extend this theorem to arbitrary normed planes. If the points $p_1, p_2, p_3$ form an equilateral set in a normed plane $(\mathbb{M}^2, \|\cdot\|)$, then the equilateral ball-polyhedron with vertices $p_1, p_2, p_3$ is a Reuleaux triangle. If $\|p_1 - p_2\| = \|p_2 - p_3\| = \|p_3 - p_1\| = \lambda$, then we denote it by $\mathcal{RT}(p_1, p_2, p_3; \lambda)$. To prove an extension of Chakerian's theorem to all normed planes we start with some lemmas.

**Lemma 2.2.1.** *Let $\mathcal{RT}(p_1, p_2, p_3; \lambda)$ be a Reuleaux triangle in a normed plane. If $x \in \mathcal{RT}(p_1, p_2, p_3; \lambda)$, then $\mathcal{RT}(p_1, p_2, p_3; \lambda) \subset \mathcal{D}(x, \lambda)$.*

*Proof.* Let $y \in \mathcal{RT}(p_1, p_2, p_3; \lambda)$. Since diam $\mathcal{RT}(p_1, p_2, p_3; \lambda) = \lambda$, we have

$$\|x - y\| \leq \lambda \Longleftrightarrow y \in \mathcal{D}(x, \lambda).$$

$\square$

**Lemma 2.2.2.** *In a normed plane $(\mathbb{M}^2, \|\cdot\|)$, let there be given three discs $\mathcal{D}(x_i, \lambda)$, $i = 1, 2, 3$, such that $x_i, x_j \in \mathcal{D}(x_k, \lambda)$ for $\{i, j, k\} = \{1, 2, 3\}$. Then $\bigcap_{i=1}^{3} \mathcal{D}(x_i, \lambda)$ contains a Reuleaux triangle of width $\lambda$.*

*Proof.* Let $\|x_1 - x_2\| = \max\{\|x_1 - x_2\|, \|x_2 - x_3\|, \|x_3 - x_1\|\}$ and $x_2' \in \mathrm{R}_{x_2}^+(x_1)$ be such that $\|x_1 - x_2'\| = \lambda$. The intersection of the circles $\mathcal{C}(x_1, \lambda)$ and $\mathcal{C}(x_2', \lambda)$ is not empty, and not all points of this intersection lie in the same half-plane with respect to the line $\mathrm{L}(x_1, x_2')$. Let

$$x_3' \in \mathcal{C}(x_1, \lambda) \cap \mathcal{C}(x_2', \lambda) \cap \mathrm{HP}_{x_3}^+(x_1, x_2').$$

If we prove that $x_3 \in \mathcal{RT}(x_1, x_2', x_3'; \lambda)$, then the statement of the lemma follows from Lemma 2.2.1. If $\|x_1 - x_2\| = \mu$, then $x_3 \in \mathcal{D}(x_1, \mu) \cap \mathcal{D}(x_2, \mu) \cap \mathrm{HS}_{x_3'}^+(x_1, x_2)$. Consider the point $x_3''$ on $\mathrm{R}_{x_3'}^+(x_1)$ such that $\|x_1 - x_3''\| = \mu$. Then $x_3'' \in \mathcal{C}(x_1, \mu) \cap \mathcal{C}(x_2, \mu)$, by Thales' theorem. Therefore

$$x_3 \in \mathrm{conv}\,\{x_1, x_2, x_3''\} \cup \mathrm{conv}\,\mathrm{arc}(x_2, x_3''; x_1) \cup \mathrm{conv}\,\mathrm{arc}(x_3'', x_1; x_2).$$

If $x_3 \in \mathrm{conv}\,\{x_1, x_2, x_3''\}$, then $x_3 \in \mathcal{RT}(x_1, x_2', x_3'; \lambda)$. Let now

$$x_3 \in \mathrm{conv}\,\mathrm{arc}(x_2, x_3''; x_1) \cup \mathrm{conv}\,\mathrm{arc}(x_3'', x_1; x_2).$$

We will prove that

$$\mathrm{conv}\,\mathrm{arc}\{x_2, x_3''; x_1\} \cup \mathrm{conv}\,\mathrm{arc}(x_3'', x_1; x_2) \subset \mathcal{D}(x_1, \lambda) \cap \mathcal{D}(x_2', \lambda) \tag{2.12}$$

We omit the case $\lambda = \mu$, which is obvious. Consider the homothety $\varphi$ mapping the circle $\mathcal{C}(x_1, \mu)$ onto the circle $\mathcal{C}(x_1, \lambda)$. Clearly, if $x$ is an arbitrary point of $\mathrm{arc}(x_2, x_3''; x_1)$, then $x' = \varphi(x)$ is a point belonging to $\mathrm{arc}(x_2', x_3'; x_1)$. Moreover, $x$ is strictly between $x_1$ and $x'$, i.e., $x \in \mathcal{D}(x_1, \lambda)$. On the other hand, the monotonicity lemma and Lemma 1.2.3 imply

$$\|x_2' - x\| \leq \max\{\|x_2' - x_1\|, \|x_2' - x'\|\} = \lambda,$$

which means that $x \in \mathcal{D}(x_2', \lambda)$. Thus we have proved that

$$\mathrm{conv}\,\mathrm{arc}(x_2, x_3''; x_1) \subset \mathcal{D}(x_1, \lambda) \cap \mathcal{D}(x_2', \lambda).$$

In order to prove that $\mathrm{arc}(x_3'', x_1; x_2)$ also belongs to $\mathcal{D}(x_1, \lambda) \cap \mathcal{D}(x_2', \lambda)$, we consider the homothety $\psi$ mapping $\mathcal{C}(x_2, \mu)$ onto $\mathcal{C}(x_2', \lambda)$. It is easy to check that the center of $\psi$ is the point

$$s = \frac{\lambda}{\lambda - \mu} x_2 - \frac{\mu}{\lambda - \mu} x_2' \tag{2.13}$$

Since $\frac{\lambda}{\lambda - \mu} > 1$, the point $s$ lies on $\mathrm{R}_{x_2'}^-(x_2)$. By (2.13) we get

$$\|s - x_2\| = \frac{\mu}{\lambda - \mu} \|x_2 - x_2'\| = \mu,$$

i.e., $s \equiv x_1$. If $y \in \mathrm{arc}(x_3'', x_1; x_2)$ and $\psi(y) = y'$, then $y' \in \mathcal{C}(x_2', \lambda)$ and $y$ lies strictly between $x_1$ and $y'$, yielding $y \in \mathcal{D}(x_1, \lambda)$. Besides this, Lemma 1.2.3 implies

$$\|x_2' - y\| \leq \max\{\|x_2' - x_1\|, \|x_2' - y'\|\} = \lambda.$$

## 2.2. On a theorem of Chakerian

Thus the inclusion (2.12) is proved and $x_3 \in \mathcal{D}(x_1, \lambda) \cap \mathcal{D}(x_2', \lambda)$. So it remains to show that if $x_3 \in \text{conv arc}(x_2, x_3''; x_1) \cup \text{conv arc}(x_3'', x_1; x_2)$, then $x_3 \in \mathcal{D}(x_3', \lambda)$. If $x_3 \in \text{conv arc}(x_2, x_3''; x_1)$ and $\text{R}_{x_3}^+(x_1) \cap C(x_1, \lambda) = \{x_3^\star\}$, then $x_3$ is strictly between $x_1$ and $x_3^\star$. Thus, again by Lemma 1.2.3 and the monotonicity lemma we have

$$\|x_3' - x_3\| \leq \max\{\|x_3' - x_1\|, \|x_3' - x_3^\star\|\} = \lambda.$$

In the same way, we can prove that $\|x_3' - x_3\| \leq \lambda$, in the case $x_3 \in \text{conv arc}(x_3'', x_1; x_2)$. □

In order to prove the main result, we also need the following generalization of Helly's theorem; see, e.g., [31, Theorem 1].

**Theorem 2.2.1. (A generalization of Helly's theorem)** *Let $P$ be a fixed compact, convex set in the plane, and $\mathfrak{F}$ be a family of compact, convex sets having the property that each three or less members of $\mathfrak{F}$ have a translate of $P$ in common. Then all the members of $\mathfrak{F}$ have a translate of $P$ in common.*

If $\psi$ is an isometry in a normed plane $(\mathbb{M}^2, \|\cdot\|)$ preserving the orientation of $(\mathbb{M}^2, \|\cdot\|)$ and $K$ is a point set in $(\mathbb{M}^2, \|\cdot\|)$, then $\psi(K)$ is called a *congruent copy* of $K$. It should be noticed that in general the only maps of $(\mathbb{M}^2, \|\cdot\|)$ that are isometries with respect to all norms are translations, reflections with respect to a point, and the identity map; see [5] and [63]. But there exist normed planes (e.g., the Euclidean plane), where the group of isometries is richer.

**Theorem 2.2.2.** *In a normed plane, let there be given a compact, convex set $P$ such that every congruent copy of $P$ can be covered by a translate of any Reuleaux triangle of Minkowskian width $\lambda$. Then each congruent copy of $P$ can be covered by a translate of any convex body of constant Minkowskian width $\lambda$.*

*Proof.* Let $K$ be an arbitrary convex body of constant Minkowskian width $\lambda$, and $x_1, x_2, x_3$ be three arbitrary points of $K$. Then $\bigcap_{i=1}^{3} \mathcal{D}(x_i, \lambda)$ contains a Reuleaux triangle $\mathcal{RT}$ of width $\lambda$, see Lemma 2.2.2. By the assumption of the theorem there exists a translate $P'$ of any congruent copy of $P$ such that $P' \subseteq \mathcal{RT}$. Therefore, by Theorem 2.2.1, all discs of radius $\lambda$ centered at $K$ have a translate of $P$ in common. Since $K$ is of constant Minkowskian width, i.e., $K = \bigcap_{x \in K} \mathcal{D}(x, \lambda)$ by Theorem 2.1.2, we conclude that $K$ contains a translate of any congruent copy of $P$, and the proof is done. □

The next statement can be obtained as an elementary corollary of Theorem 2.2.2.

**Corollary 2.2.1.** *In a normed plane, let there be given a finite point set $P$ such that every congruent copy of $P$ can be covered by a translate of any Reuleaux triangle of Minkowskian width $\lambda$. Then any congruent copy of the convex hull of $P$ can be covered by a translate of any convex body of constant Minkowskian width $\lambda$.*

## 2.3 Further characterizations of bodies of constant width

The notion of a *normal* to a convex body was introduced by Eggleston (see [35], [36], and the survey [32]). It is a very useful notion, e.g. for characterizations of convex bodies of constant width in Euclidean space, as well as in any finite-dimensional real Banach space. Following this approach, we introduce the notion of *affine orthogonality* with respect to a convex body. If the convex body is a circular disc, then our definition coincides with the usual Euclidean orthogonality. But for bodies of constant width we have coincidence with the notion of normals of Eggleston. The advantage of our approach is that via this notion we are able to characterize not only bodies of constant width but also other classes of convex bodies, such as centrally symmetric ones, ellipses, bodies whose boundary is a Radon curve, etc. Moreover, we contribute to the solution of the following problem posed by V. Soltan: to extend the characterization of bodies of constant width in the Euclidean plane given by Makai Jr. and Martini in [59] to normed planes by replacing Euclidean orthogonality by Birkhoff orthogonality. Note that, due to the counterexample constructed in [1], the mentioned characterization cannot be extended to all normed planes based on Birkhoff orthogonality. But if we replace Euclidean orthogonality by affine orthogonality (see our Theorem 2.3.3), then we obtain such an extension. It should be noticed that in contrast to Eggleston's definition, our definition does not need a metric. In other words, our considerations take place in an arbitrary affine space.

### 2.3.1 Affine orthogonality

We start with the definition of affine orthogonality in the planar case. Let $[p_1, p_2]$ and $[q_1, q_2]$ be two chords of a convex body $K$ in $\mathbb{M}^2$, and let $P_1$ be the line through $p_1$ parallel to $L(q_1, q_2)$. We say that $[p_1, p_2]$ *is affinely orthogonal to* $[q_1, q_2]$ *through* $p_1$, denoted by

$$[p_1, p_2] \dashv_{p_1} [q_1, q_2],$$

if one of the following two conditions holds:

a) $P_1$ supports $K$ at $p_1$, and also the line through $p_2$ parallel to $P_1$ supports $K$.

b) $P_1 \cap \operatorname{bd} K = \{p_1, p_1'\}$, and $[p_1', p_2]$ is an affine diameter of $K$.

If $K$ is a Euclidean disc, then the so-defined relation coincides with the usual Euclidean orthogonality. It is also clear that, in general, affine orthogonality is not symmetric with respect to the chords $[p_1, p_2]$ and $[q_1, q_2]$. It is also not symmetric with respect to the points $p_1$ and $p_2$. As we shall see later, these properties of symmetry will characterize special types of convex bodies. On the other hand, affine orthogonality is obviously symmetric with respect to $q_1$ and $q_2$.

## 2.3. FURTHER CHARACTERIZATIONS OF BODIES OF CONSTANT WIDTH

The next proposition follows directly from the definition above. Property (1) shows that the position of the second chord $[q_1, q_2]$ does not matter: of importance is only the (non-oriented) direction of $[q_1, q_2]$. Property (2) recalls what happens if $K$ is a Euclidean disc: if $p$, $q$ and $r$ are points in bd $K$ and $[q, r]$ is a diameter, then $[p, q]$ is orthogonal to $[p, r]$. This is, in fact, Thales' theorem and was motivating for the definition of affinely orthogonal chords.

**Proposition 2.3.1.** *Let $K$ be a convex body in $\mathbb{M}^2$.*

(1) *If $[p_1, p_2] \dashv_{p_1} [q_1, q_2]$, then for any chord $[q_1', q_2']$ of $K$ that is parallel to $[q_1, q_2]$ the relation $[p_1, p_2] \dashv_{p_1} [q_1', q_2']$ holds.*

(2) *Let $[p_2, q_2]$ be an affine diameter of $K$ and let the point $p_1 \in $ bd $K$ be such that the segment $[p_1, q_2]$ does not belong to the boundary of $K$. Then $[p_1, p_2] \dashv_{p_1} [p_1, q_2]$.*

Let now $K$ be a convex body in $\mathbb{M}^d$ ($d \geq 3$). The intersection of $K$ with a two-dimensional flat $\alpha$ is a planar convex body (*plane section*) that will be denoted by $K_\alpha$. Let $[p_1, p_2]$ and $[q_1, q_2]$ be two intersecting chords of $K$ and let $\alpha$ be the 2-flat that contains them. We say that $[p_1, p_2]$ is affinely orthogonal through $p_1$ to $[q_1, q_2]$, denoted again $[p_1, p_2] \dashv_{p_1} [q_1, q_2]$, if $[p_1, p_2]$ is affinely orthogonal through $p_1$ to $[q_1, q_2]$ with respect to $K_\alpha$. For defining affine orthogonality it is not necessary that $[p_1, p_2]$ and $[q_1, q_2]$ intersect, but sufficient that there exists a 2-flat containing both chords. If $\alpha$ passes through the origin, we call $K_\alpha$ a *main plane section of $K$*, and if two chords of $K$ determine a 2-flat passing through the origin, then we call them *main chords*.

### 2.3.2 Characterizations of bodies of constant width via affine orthogonality

Let $(\mathbb{M}^d, \|\cdot\|)$ be a normed space, and let $K \subset \mathbb{M}^d$ be a convex body. A chord $[p, q]$ of $K$ is called a *normal of $K$ at $p$* if $[p, q]$ is Birkhoff orthogonal to a supporting hyperplane of $K$ at $p$. This notion was introduced by Eggleston (see [35, p. 166]). Via the notion of normals of a convex body Eggleston gave the following characterization of bodies of constant width which is not only of interest in itself. In the Euclidean subcase this characterization forms the basis for the usual definition of space curves of constant width and the generalization of that concept to transnormal manifolds embedded in Euclidean spaces.

**Theorem 2.3.1.** ([32, (VI')]) *In a strictly convex and smooth normed space, a convex body $K$ is of constant width if and only if every chord $[p, q]$ of $K$ that is a normal of $K$ at $p$ is also a normal of $K$ at $q$.*

Our next theorem characterizes bodies of constant width by relating affinely orthogonal chords to normal chords.

**Theorem 2.3.2.** *Let $K$ be a convex body in a strictly convex and smooth normed plane. The following properties are equivalent:*

(i) *$K$ is of constant width.*

(ii) *If $[p_1,p_2]$ is a normal chord of $K$ at $p_1$ and $[p_1,p_2]$ is Birkhoff orthogonal to the chord $[q_1,q_2]$, then $[p_1,p_2] \dashv_{p_1} [q_1,q_2]$.*

*Proof.* (i)$\Rightarrow$(ii) Assume that $[p_1,p_2]$ is a normal chord of $K$ at $p_1$. Then $[p_1,p_2]$ is Birkhoff orthogonal to a line $L$ that supports $K$ at $p_1$. By Theorem 2.3.1, the line through $p_2$ parallel to $L$ supports $K$ at $p_2$. If $[p_1,p_2]$ is Birkhoff orthogonal to the chord $[q_1,q_2]$, then $[q_1,q_2]$ is parallel to $L$, from which it follows that $[p_1,p_2] \dashv_{p_1} [q_1,q_2]$.

(ii)$\Rightarrow$(i) Let $[p_1,p_2]$ be a chord of $K$ that is a normal of $K$ at $p_1$, and let $L$ be a supporting line of $K$ at $p_1$ such that $[p_1,p_2]$ is Birkhoff orthogonal to $L$. Let $[q_1,q_2]$ be any chord parallel to $L$. Then $[p_1,p_2]$ is Birkhoff orthogonal to $[q_1,q_2]$, and by (ii) we have $[p_1,p_2] \dashv_{p_1} [q_1,q_2]$, which implies that the line through $p_2$ parallel to $L$ supports $K$ at $p_2$. Thus $[p_1,p_2]$ is also a normal of $K$ at $p_2$ and, by Theorem 2.3.1, $K$ is of constant width. $\square$

In [59] Martini and Makai Jr. gave the following characterization of bodies of constant width in the Euclidean plane: a convex body of diameter 1 in $\mathbb{E}^2$ is of constant width 1 if and only if any two perpendicular chords of it have total length greater than or equal to 1. V. Soltan posed the question of extending this characterization to normed planes by replacing the usual Euclidean orthogonality by Birkhoff orthogonality. But as the counterexample constructed in [1] shows, in general this cannot be done. Now we prove that such an extension is possible if Euclidean orthogonality is replaced by affine orthogonality. For that purpose we need the following lemma.

**Lemma 2.3.1.** [92, **Property 3.2**] *Let $K \subset \mathbb{M}^d$ be a convex body. Then for any line $L$ there exist $q_1, q_2 \in \mathrm{bd}\, K$ such that $[q_1, q_2]$ is an affine diameter of $K$ and $\mathrm{L}(q_1, q_2)$ is parallel to $L$.*

Let $[q_1,q_2]$ be a chord of a convex body $K \subset \mathbb{M}^2$, and let $p \in \mathrm{bd}\, K$. We say that *$p$ is in the neighbourhood of* $[q_1,q_2]$ if there exists an affine diameter $[q_1',q_2']$ which is parallel to $[q_1,q_2]$ such that $[p,q_1'] \cap [q_1,q_2] \neq \emptyset$. And we say that a convex body $K$ in a normed plane has the *affine orthogonality property* if for any two intersecting chords $[p_1,p_2]$ and $[q_1,q_2]$, with $p_1$ in the neighbourhood of $[q_1,q_2]$ and $[p_1,p_2] \dashv_{p_1} [q_1,q_2]$, the inequality $\|p_1-p_2\| + \|q_1-q_2\| \geq \mathrm{diam}\, K$ holds.

**Theorem 2.3.3.** *A convex body $K$ in a normed plane is of constant width if and only if it has the affine orthogonality property.*

## 2.3. FURTHER CHARACTERIZATIONS OF BODIES OF CONSTANT WIDTH

*Proof.* Assume that $K$ is of constant width, and let $[p_1, p_2]$ and $[q_1, q_2]$ be two intersecting chords of $K$ such that $[p_1, p_2] \dashv_{p_1} [q_1, q_2]$ and $p_1$ is in the neighbourhood of $[q_1, q_2]$. Let $P_1$ be the line through $p_1$ and parallel to $[q_1, q_2]$. Then there exists $p_1' \in P_1 \cap \mathrm{bd}\, K$ (it is possible that $p_1' = p_1$) such that $[p_1', p_2]$ is an affine diameter. Since $p_1$ is in the neighbourhood of $[q_1, q_2]$, we have that $\|p_1 - p_1'\| \leq \|q_1 - q_2\|$, and then

$$\mathrm{diam}\, K = \|p_2 - p_1'\| \leq \|p_2 - p_1\| + \|p_1 - p_1'\| \leq \|p_1 - p_2\| + \|q_1 - q_2\|.$$

Conversely, assume that $K$ has the affine orthogonality property and that there exists an affine diameter $[x, y]$ with $\|x - y\| < \mathrm{diam}\, K$. Let $P$ be a supporting line of $K$ at $x$, and let us first assume that $P$ touches $\mathrm{bd}\, K$ only at $x$. Then there exists a chord $[q_1, q_2]$ of $K$ parallel to $P$ and such that $\|q_1 - q_2\| < \mathrm{diam}\, K - \|x - y\|$, with $x$ in the neighbourhood of $[q_1, q_2]$. But then $[x, y] \dashv_x [q_1, q_2]$ and $\|q_1 - q_2\| + \|x - y\| < \mathrm{diam}\, K$, which contradicts the affine orthogonality property. On the other hand, if $P$ touches $\mathrm{bd}\, K$ at a segment that contains $x$ then, taking a point $x'$ in that segment such that $\|x - x'\| < \mathrm{diam}\, K - \|x - y\|$, we get also a contradiction, since $[x, y] \dashv_x [x, x']$. □

Using the following result from geometric tomography, we extend the "$\Longleftarrow$" part of Theorem 2.3.3 to Euclidean space of dimension $d \geq 3$.

**Theorem 2.3.4.** [39, Corollary 7.1.15] *Let $K$ be a convex body in an Euclidean space $\mathbb{E}^d$ containing the origin in its interior. If any main plane section of $K$ is of constant width, then $K$ is also of constant width.*

**Theorem 2.3.5.** *Let $K$ be a convex body in an Euclidean space $\mathbb{E}^d$, where $d \geq 3$. If $K$ contains the origin in its interior and for any two main chords $[p_1, p_2]$ and $[q_1, q_2]$ with $p_1$ in the neighbourhood of $[q_1, q_2]$ and $[p_1, p_2] \dashv_{p_1} [q_1, q_2]$, the inequality $\|p_1 - p_2\| + \|q_1 - q_2\| \geq \mathrm{diam}\, K$ holds, then $K$ is of constant width.*

*Proof.* It follows directly from Theorem 2.3.3 and Theorem 2.3.4. □

Due to further results from geometric tomography we are also able to describe what happens when all main plane sections of a convex body are of constant width.

**Lemma 2.3.2.** [39, Lemma 7.1.14] *If $K$ is a convex body in Euclidean space $\mathbb{E}^d$, $d \geq 3$, containing the origin in its interior and for any hyperplane $\sigma$ through the origin $K \cap \sigma$ is of constant width, then there is a diameter $D$ of $K$ containing the origin. Furthermore, if a main plane section contains $D$, then each normal of $K \cap \sigma$ is a diameter of $K$.*

If a 2-flat $\alpha$ contains the diameter $D$ from Lemma 2.3.2, we call the plane section $K_\alpha$ a *maximal plane section*. Since $D$ passes through the origin, any maximal plane section is also a main plane section.

**Theorem 2.3.6.** *Let $K$ be a convex body in Euclidean space $\mathbb{E}^3$ containing the origin in its interior, and any main plane section of $K$ be of constant width. Let $K_\alpha$ be a maximal plane section.*

1. *Then for any two chords $[p_1, p_2]$ and $[q_1, q_2]$ of $K_\alpha$ with $p_1$ in the neighbourhood of $[q_1, q_2]$ and $[p_1, p_2] \dashv_{p_1} [q_1, q_2]$, the inequality $\|p_1 - p_2\| + \|q_1 - q_2\| \geq \operatorname{diam} K$ holds.*

2. *Let $[p_2, q_2]$ be a normal of $K_\alpha$ and let the point $p \in \operatorname{bd} K_\alpha$ be such that the segment $[p, q_2]$ does not belong to the boundary of $K_\alpha$. Then $\|p_2 - p\| + \|p - q_2\| \geq \operatorname{diam} K$.*

*Proof.* The first claim follows from Lemma 2.3.2 and Theorem 2.3.3. For the second one we note that any normal of $K_\alpha$ is also a double normal. But every double normal is an affine diameter, and Proposition 2.3.1 yields $[p, p_2] \dashv_p [p, q_2]$. Thus claim 2. is a consequence of 1. □

**Remark 2.3.1.** *The mentioned characterization of Makai Jr. and Martini is extended to Euclidean space of dimension $d \geq 3$ only in the one direction. Namely if for a convex body $K$ of diameter 1 any $d$ mutually perpendicular chords of it have total length greater than or equal to 1 then $K$ is of constant width; see [59]. But it is still unknown even in dimension 3 whether the converse holds, i.e., whether any convex body of constant width 1 has this property. Replacing the usual orthogonality in Euclidean spaces by an affine orthogonality our Theorem 2.3.6 solves this problem for a big class of convex bodies of constant width in $\mathbb{E}^3$. According to Theorem 2.3.4, this is due to the fact that if any main plane section of a convex body is of constant width, then this body is also of constant width.*

### 2.3.3 Applications of affine orthogonality for characterizations of further classes of special convex bodies

As we already mentioned, the advance of the definition of affine orthogonality is that we are able to characterize not only bodies of constant width, but also centrally symmetric bodies, Radon curves, and ellipses.

We start with a relation between the notion of affine orthogonality and the class of centrally symmetric convex bodies. If $K$ is centrally symmetric and $p \in K$, then we denote by $\bar{p}$ the point opposite to $p$ with respect to the center of $K$. Thus, if $K$ is centrally symmetric and $p \in \operatorname{bd} K$, then $[p, \bar{p}]$ is an affine diameter. But if in the boundary of a centrally symmetric convex body $K$ there is a segment that contains $p_1$ or $p_2$, then $[p_1, p_2]$ can be an affine diameter which does not pass through the center of $K$. Example (A) in Figure 2.4 shows that, in general, $[p_1, p_2] \dashv_{p_1} [q_1, q_2]$ does not imply $[p_1, p_2] \dashv_{p_2} [q_1, q_2]$.

**Theorem 2.3.7.** *For a convex body $K \subset \mathbb{M}^2$, the following properties are equivalent:*

2.3. FURTHER CHARACTERIZATIONS OF BODIES OF CONSTANT WIDTH      41

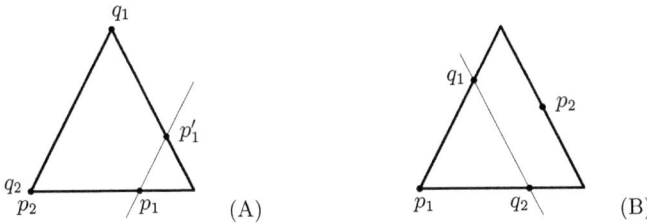

Figure 2.4: (A) $[p_1, p_2] \dashv_{p_1} [q_1, q_2]$, but $[p_1, p_2] \not\dashv_{p_2} [q_1, q_2]$, (B) $[p_1, p_2] \dashv_{p_1, p_2} [q_1, q_2]$.

(i) $K$ is centrally symmetric and strictly convex.

(ii) If $[p_1, p_2]$ and $[q_1, q_2]$ are two chords of $K$ and $[p_1, p_2] \dashv_{p_1} [q_1, q_2]$, then $[p_1, p_2] \dashv_{p_2} [q_1, q_2]$.

*Proof.* (i)⇒(ii) Let $x$ be the center of $K$ and assume that $[p_1, p_2] \dashv_{p_1} [q_1, q_2]$. Let $P_1$ be the line through $p_1$ parallel to $L(q_1, q_2)$. If $P_1$ supports $K$ at $p_1$, then trivially $[p_1, p_2] \dashv_{p_2} [q_1, q_2]$. Let now $P_1$ not support $K$ at $p_1$, and $P_1 \cap \operatorname{bd} K = \{p_1, p_1'\}$. Then $[p_1', p_2]$ is an affine diameter, and since $K$ is strictly convex, $p_1'$ and $p_2$ are opposite with respect to $x$, i.e., $p_1' = \bar{p}_2$. Since $L(\bar{p}_1, p_2)$ is parallel to $L(p_1, \bar{p}_2) = P_1$ and $[p_1, \bar{p}_1]$ is an affine diameter, we get $[p_1, p_2] \dashv_{p_2} [q_1, q_2]$.

(ii)⇒(i) Let us show first that property (ii) implies that $K$ is strictly convex. Assume, on the contrary, that bd $K$ contains a segment $[p, q]$, and let $p_1$ and $p_2$ be two different points in the interior of that segment. Let $q_2 \in \operatorname{bd} K$ be such that $[p_2, q_2]$ (and then also $[p_1, q_2]$) is an affine diameter. Then we have that $[p_1, p_2] \dashv_{p_1} [p_1, q_2]$. By (ii) also $[p_1, p_2] \dashv_{p_2} [p_1, q_2]$. Since the line $Q$ through $p_2$ and parallel to $L(p_1, q_2)$ does not support $K$, we have that $Q$ cuts bd $K$ in a point $q_2'$ such that $[p_2, q_2']$ is an affine diameter and $L(p_1, p_2)$ is parallel to $L(q_2, q_2')$. But this implies that $[q_2, q_2']$ is a segment contained in bd $K$ and having the same length as $[p_1, p_2]$. Interchanging the roles of $p_1$ and $p_2$, the above argument shows that there exists another point $q_2''$ in bd $K$ such that $q_2$ is the midpoint of $[q_2', q_2'']$, and thus bd $K$ contains a segment having the double length of $[p_1, p_2]$. Since $p_1$ and $p_2$ are arbitrary points in the interior of $[p, q]$, we have proved that if bd $K$ contains a segment, then it contains another segment of double length, which is an absurdity.

Let us now show that $K$ has the following property:

(∗) If $[p, q]$ is an affine diameter and $P$ is a line that supports $K$ at $p$, then the line $Q$ through $q$ and parallel to $P$ also supports $K$.

Indeed, assume that $Q \cap \operatorname{bd} K = \{q, q'\}$, where $q \neq q'$. Then $[p, q'] \dashv_{q'} [q, q']$, and by (ii)

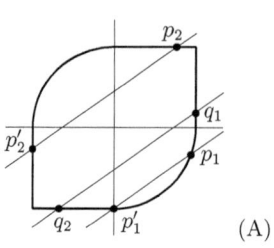 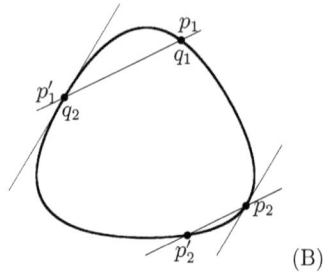

Figure 2.5: (A) Centrally symmetric but not strictly convex, (B) Strictly convex but not centrally symmetric. In both cases $[p_1, p_2] \dashv_{p_1} [q_1, q_2]$ but $[p_1, p_2] \not\dashv_{p_2} [q_1, q_2]$.

we have $[p, q'] \dashv_p [q, q']$. But this implies that $Q$ supports $K$ at $q$ (recall (a) in the definition of affine orthogonality), which is impossible because $K$ is strictly convex.

Fix now an affine diameter $[p_2, q_2]$ of $K$. We shall see that $K$ is centered at the midpoint of $[p_2, q_2]$. Denote by $P_2$ and $Q_2$ the parallel supporting lines at $p_2$ and $q_2$, respectively. Let $p_1$ be an arbitrary point in bd $K$, different from $p_2$ and $q_2$. By Proposition 2.3.1 we have that $[p_1, p_2] \dashv_{p_1} [p_1, q_2]$, and then by (ii) we have $[p_1, p_2] \dashv_{p_2} [p_1, q_2]$. This, together with property (∗), implies that the line $P_2'$ through $p_2$ and parallel to $[p_1, q_2]$ cuts bd $K$ in a point $p_2' \neq p_2$, with $[p_1, p_2']$ being an affine diameter. Moreover, we have $[p_2', p_2] \dashv_{p_2'} [p_2', q_2]$, and by (ii) we get $[p_2', p_2] \dashv_{p_2} [p_2', q_2]$. Again by (∗), the line $P_2''$ through $p_2$ and parallel to $[p_2', q_2]$ cuts bd $K$ in a point $p_2'' \neq p_2$, and $[p_2', p_2'']$ is an affine diameter. Since $[p_1, p_2']$ and $[p_2', p_2'']$ are both affine diameters, property (∗) and the strict convexity of $K$ imply that $p_2'' = p_1$. Hence the points $p_1, p_2, p_2'$ and $q_2$ form a parallelogram, and $p_2'$ is the point symmetric to $p_1$ with respect to the midpoint of $[p_2, q_2]$. □

In Figure 2.5 we see examples confirming that if in Theorem 2.3.7 property (i) fails, then also (ii) fails. On the other hand, as Proposition 2.3.2 below shows that if the chord $[p_1, p_2]$ has a special position, then strict convexity is not necessary to obtain property (ii).

**Proposition 2.3.2.** *Let $p, q_1, q_2$ be three different points of a centrally symmetric convex body $\mathcal{K}$. If $[p, \bar{p}] \dashv_p [q_1, q_2]$, then the line $P$ through $p$ and parallel to $[q_1, q_2]$ supports $\mathcal{K}$ at $p$, yielding $[p, \bar{p}] \dashv_{\bar{p}} [q_1, q_2]$.*

*Proof.* Assume that $[p, \bar{p}] \dashv_p [q_1, q_2]$, but $P \cap$ bd $K = \{p, p'\}$, where $p \neq p'$. Then $\mathcal{K}$ has supporting lines at $\bar{p}$ and at $p'$ that are parallel, and by the symmetry of $\mathcal{K}$ these lines are also parallel to a supporting line at $p$ that, by the convexity of $\mathcal{K}$, coincides with $P$, a contradiction. □

## 2.3. FURTHER CHARACTERIZATIONS OF BODIES OF CONSTANT WIDTH

In view of the situations described in Theorem 2.3.7 and in Proposition 2.3.2, if $[p_1, p_2] \dashv_{p_1} [q_1, q_2]$ and $[p_1, p_2] \dashv_{p_2} [q_1, q_2]$, we simply write $[p_1, p_2] \dashv [q_1, q_2]$.

**Remark 2.3.2.** *Let $\mathcal{K}$ be a centrally symmetric convex body in $\mathbb{M}^2$, taken as the unit ball of a norm. Then from Proposition 2.3.2 it follows that in this case $[p, \bar{p}] \dashv [q_1, q_2]$ if and only if $p - \bar{p}$ is Birkhoff orthogonal to $q_1 - q_2$.*

To extend Theorem 2.3.7 to higher dimensions, we need the following theorem known as the *false center theorem*.

**Theorem 2.3.8.** [39, Corollary 7.1.10] *Let $K$ be a convex body in $\mathbb{E}^d$, $d \geq 3$. If any main plane section of $K$ is centrally symmetric, then $K$ is centrally symmetric about the origin or an ellipsoid.*

From this theorem and Theorem 2.3.7 we have immediately

**Theorem 2.3.9.** *Let $K$ be a convex body in an Euclidean space $\mathbb{E}^d$, $d \geq 3$.*

1. *Assume that $K$ is strictly convex and symmetric about the origin. If $[p_1, p_2]$ and $[q_1, q_2]$ are two main chords with $[p_1, p_2] \dashv_{p_1} [q_1, q_2]$, then $[p_1, p_2] \dashv_{p_2} [q_1, q_2]$.*

2. *Assume that for any two main chords $[p_1, p_2]$ and $[q_1, q_2]$ the implication $[p_1, p_2] \dashv_{p_1} [q_1, q_2] \implies [p_1, p_2] \dashv_{p_2} [q_1, q_2]$ holds. Then $K$ is strictly convex and symmetric about the origin or an ellipsoid.*

The next class of convex bodies that can be described via affine orthogonality is the class of those convex bodies in $(\mathbb{M}^2, \|\cdot\|)$ whose boundary is a Radon curve. Radon curves were introduced by Radon [83] in 1916 and independently rediscovered by Birkhoff [19]. A centrally symmetric, closed, convex curve $\mathcal{C}$ is called a *Radon curve* if it has the following property:

For $p \in \mathcal{C}$, let $P$ be a supporting line of $\mathcal{C}$ at $p$ and assume that the line parallel to $P$ through the center of $\mathcal{C}$ intersects $\mathcal{C}$ at $q$ and $\bar{q}$. Then the line through $q$ parallel to $\mathrm{L}(p, \bar{p})$ supports $\mathcal{C}$.

Any Radon curve centered at the origin defines a norm whose properties are "almost Euclidean". In fact, the boundary of a centrally symmetric convex body $K$ in $\mathbb{M}^2$ is a Radon curve if and only if Birkhoff orthogonality with respect to the induced norm is symmetric. For $d \geq 3$ the symmetry of the Birkhoff orthogonality characterizes Euclidean spaces. In other words, a centrally symmetric convex body in $\mathbb{M}^d$ ($d \geq 3$) is an ellipsoid if and only if the boundaries of its two-dimensional sections through the center of symmetry are Radon curves. This characterization was obtained in gradual stages by G. Birkhoff [19], R. C. James [51, 52] and M. M. Day [33]. For further properties of Radon curves we refer, e.g., to [99, § 4.7], [72], and [71].

**Theorem 2.3.10.** *For a centrally symmetric convex body $K \subset \mathbb{M}^2$, the following properties are equivalent:*

(i) *The boundary of $K$ is a Radon curve.*

(ii) *If $p, q \in \mathrm{bd}\, K$ with $[p, \bar{p}] \dashv [q, \bar{q}]$, then $[q, \bar{q}] \dashv [p, \bar{p}]$.*

*Proof.* (i)$\Rightarrow$(ii) This implication follows directly from Remark 2.3.2. (ii)$\Rightarrow$(i) Let $p \in \mathrm{bd}\, K$, let the line $P$ support $K$ at $p$, and $q \in \mathrm{bd}\, K$ be such that $[q, \bar{q}]$ is parallel to $P$. Then $[p, \bar{p}] \dashv [q, \bar{q}]$, and therefore $[q, \bar{q}] \dashv [p, \bar{p}]$. By Proposition 2.3.2 we obtain that the line $Q$ through $q$ and parallel to $[p, \bar{p}]$ supports $K$ at $q$, which implies that $\mathrm{bd}\, K$ is a Radon curve. $\square$

Theorem 2.3.10 above shows that the fact that affine orthogonality is symmetric over a particular class of chords of $K$ is characteristic for Radon curves. The next theorem shows that if this symmetry is extended to a wider class of affinely orthogonal chords, then it is even characteristic for ellipses.

**Theorem 2.3.11.** *For a centrally symmetric convex body $K \subset \mathbb{M}^2$, the following properties are equivalent:*

(i) *The boundary of $K$ is an ellipse.*

(ii) *If $p, q_1, q_2 \in \mathrm{bd}\, K$ and $[p, \bar{p}] \dashv [q_1, q_2]$, then $[q_1, q_2] \dashv_{q_1} [p, \bar{p}]$ or $[q_1, q_2] \dashv_{q_2} [p, \bar{p}]$.*

*Proof.* (i)$\Rightarrow$(ii) This is evident. (ii)$\Rightarrow$(i) First we show that $K$ is strictly convex. Assume the contrary, namely that $[s, t]$ is a segment contained in $\mathrm{bd}\, K$ and that there is no larger segment containing it. Assume, without loss of generality, that the center of $K$ is the origin of $\mathbb{M}^2$. Let $p = \frac{1}{4}s + \frac{3}{4}t$, $q_1 = \frac{4}{5}\bar{s} + \frac{1}{5}\bar{t}$, and $q_2 = \frac{5}{6}\bar{s} + \frac{1}{6}\bar{t}$. Then $p \in [s, t]$, $q_1, q_2 \in [\bar{s}, \bar{t}]$, and $[p, \bar{p}] \dashv [q_1, q_2]$. If $[q_1, q_2] \dashv_{q_1} [p, \bar{p}]$, then $p + q_1 - \bar{p} = 2p + q_1 = \frac{-3}{10}s + \frac{13}{10}t \in \mathrm{bd}\, K$, which implies $[s, t] \subsetneq [s, \frac{-3}{10}s + \frac{13}{10}t] \subset \mathrm{bd}\, K$, contradicting the hypothesis. If $[q_1, q_2] \dashv_{q_2} [p, \bar{p}]$, we obtain a similar result. Thus, Theorem 2.3.11, property (ii), reads as: *If $p, q_1, q_2 \in \mathrm{bd}\, K$ and $[p, \bar{p}] \dashv [q_1, q_2]$, then $[q_1, q_2] \dashv [p, \bar{p}]$.*

Now we shall see that the midpoints of every family of parallel chords lie in a line, which is a well known characterization of ellipses; see, e.g., [51]. Let $q \in \mathrm{bd}\, K$, and let $P$ be a line parallel to $[q, \bar{q}]$ that supports $K$ at a point, say $p$. Then $[p, \bar{p}] \dashv [q_1, q_2]$ for any chord $[q_1, q_2]$ parallel to $[q, \bar{q}]$. By (ii), $[q_1, q_2] \dashv [p, \bar{p}]$. Let $Q_i$, $i = 1, 2$, denote the lines parallel to $[p, \bar{p}]$ and passing through $q_i$. By the definition of affine orthogonality it follows that if $Q_1$ supports $K$, then $Q_2$ also supports $K$, and since $K$ is strictly convex we get $q_2 = \bar{q}_1$. This implies that the midpoint of $[q_1, q_2]$ is the center of $K$, thus lying in $[p, \bar{p}]$. On the other hand, if $Q_1 \cap \mathrm{bd}\, K = \{q_1, q'_1\}$, then $[q'_1, q_2]$ is an affine diameter and its midpoint is again the center of $K$. Since $Q_1$ is parallel to $[p, \bar{p}]$, this chord cuts $[q_1, q_2]$ in its midpoint. $\square$

## 2.3. FURTHER CHARACTERIZATIONS OF BODIES OF CONSTANT WIDTH

**Theorem 2.3.12.** *For a centrally symmetric convex body $K \subset \mathbb{M}^2$, the following properties are equivalent:*

(i) *The boundary of $K$ is a circular disc.*

(ii) *For $p, q_1, q_2 \in \operatorname{bd} K$ the relation $[p, \bar{p}] \dashv [q_1, q_2]$ implies that then $[p, \bar{p}]$ is orthogonal to $[q_1, q_2]$ in the Euclidean sense.*

*Proof.* (i)$\Rightarrow$(ii) This is evident. (ii)$\Rightarrow$(i) First we show that $K$ is smooth. Assume, on the contrary, that there are two different supporting lines at a point $p$ of bd $K$, say $L_1$ and $L_2$. Let $q_1, q_2 \in \operatorname{bd} K$ be such that $[q_1, \bar{q}_1]$ is parallel to $L_1$ and $[q_2, \bar{q}_2]$ is parallel to $L_2$. Then $[p, \bar{p}] \dashv [q_1, \bar{q}_1]$ and $[p, \bar{p}] \dashv [q_2, \bar{q}_2]$, which implies that $[p, \bar{p}]$ is orthogonal to $[q_1, \bar{q}_1]$ and to $[q_2, \bar{q}_2]$ in the Euclidean sense, which is absurd. Without loss of generality we can assume that $K$ is centered at the origin, and since it is smooth we can parameterize bd $K$ via a function

$$\theta \in [0, 2\pi] \to x(\theta) = \rho(\theta)(\cos\theta, \sin\theta) \in \operatorname{bd} K,$$

where $\rho(\theta)$ is a positive differentiable function. Then, for each $\theta \in [0, 2\pi]$, the line through $x(\theta)$ parallel to $x'(\theta)$ supports $K$ at $x(\theta)$, which implies that $[x(\theta), -x(\theta)]$ is affinely orthogonal (and then also orthogonal in the Euclidean sense) to that line. Therefore, the scalar product $x(\theta) \cdot x'(\theta)$ is zero, and then

$$0 = \rho(\theta)(\cos\theta, \sin\theta) \cdot (\rho'(\theta)\cos\theta - \rho(\theta)\sin\theta, \rho'(\theta)\sin\theta + \rho(\theta)\cos\theta) = \rho(\theta)\rho'(\theta),$$

which implies that $\rho'(\theta) = 0$ for $\theta \in [0, 2\pi]$. Consequently, bd $K$ is the circle with center zero and of radius $\rho(0)$. $\square$

The next theorem from geometric tomography gives what is necessary to extend Theorem 2.3.11 and Theorem 2.3.12 to higher dimensions.

**Theorem 2.3.13.** [39, **Theorem 7.1.5 and Corollary 7.1.4**] *Let $K$ be a convex body in $\mathbb{E}^d$, $d \geq 3$.*

1. *If $K$ contains the origin in its relative interior and any main plane section of $K$ is an ellipse, then $K$ is an ellipsoid.*

2. *If any main plane section is a ball, then $K$ is also a ball.*

The above theorem, Theorem 2.3.11, and Theorem 2.3.12 yield

**Theorem 2.3.14.** *Let $K$ be a convex body in $\mathbb{E}^d$, $d \geq 3$.*

1. *If $K$ is a centrally symmetric strictly convex body whose center is the origin, then $K$ is an ellipsoid if and only if for any two main chords $[p_1, p_2]$ and $[q_1, q_2]$ with $[p_1, p_2] \dashv [q_1, q_2]$ the relation $[q_1, q_2] \dashv [p_1, p_2]$ holds.*

2. If $K$ is centrally symmetric strictly convex body whose center is the origin, then $K$ is a ball if and only if for any two main chords $[p_1,p_2]$ and $[q_1,q_2]$ with $[p_1,p_2] \dashv [q_1,q_2]$ it follows that $[p_1,p_2]$ and $[q_1,q_2]$ are orthogonal in the Euclidean sense.

# Chapter 3

# Kissing spheres. Covering and packing balls

In this chapter we solve kissing, covering, and packing problems related to spheres and balls in normed spaces. The results included here are from the papers [95], [64], [66], [94], and [2]. Lemma 3.2.2, Theorem 3.3.2, Theorem 3.4.1 are proved by Martini and Proposition 3.5.3, Theorem 3.5.1, (iii) by Alonso.

## 3.1 Kissing spheres

In a normed space $(\mathbb{M}^d, \|\cdot\|)$, let $D_1$ and $D_2$ be two balls and $S_i = \text{bd } D_i$, $i = 1, 2$. The spheres $S_1$ and $S_2$ are called *kissing spheres* if $S_1 \cap S_2 \neq \emptyset$ and one of the following situations holds:

1. $D_1$ and $D_2$ are non-overlapping balls;

2. $D_1 \subsetneq D_2$;

3. $D_2 \subsetneq D_1$.

It is clear that two kissing spheres in a two-dimensional strictly convex normed plane have exactly one point in common, i.e., they are touching circles as defined in Chapter 1. If a hyperplane supports a sphere, we also say that these sets are *kissing*. Two hyperplanes are called *kissing hyperplanes* if they are parallel. Let $\mathfrak{F}$ be a family of spheres and hyperplanes. Such a family is said to be a *kissing family* if any two members of it are kissing. It is our aim to find for a given family (not necessary a kissing family) of spheres and hyperplanes in $(\mathbb{M}^d, \|\cdot\|)$, *all* spheres which kiss any member of this family. If the given family consists of $d+1$ spheres and hyperplanes, then for the Euclidean subcase we obtain the famous *Apollonius problem*. Note that the Apollonius problem is a subject of many investigations in incidence geometries.

48  CHAPTER 3. KISSING SPHERES. COVERINGS AND PACKINGS BY BALLS

Now we present the first investigations on this problem in normed planes. For that reason we prove that strictly convex, smooth normed plane can be viewed as an incidence structures, more precisely, they are topological Möbius planes. Note that the term "Möbius plane" is used according to Benz' terminology. Möbius planes in this sense generalize the classical Möbius planes, i.e., an affine plane, where the set of all lines and circles is invariant with respect to the group of Möbius transformations.

### 3.1.1 Strictly convex, smooth normed planes as topological Möbius planes

We start with the definition of Möbius planes. Let $\mathfrak{P}$ be a set, and $\mathfrak{C}$ be a set of subsets of $\mathfrak{P}$. The elements of $\mathfrak{P}$ are said to be *generalized points*, and the elements of $\mathfrak{C}$ are *generalized circles*. The incidence structure $(\mathfrak{P}, \mathfrak{C})$ is called *Möbius plane* if it satisfies the following axioms:

**M1.** For any three distinct generalized points $x, y, z$ there exists a unique generalized circle incident with them.

**M2.** For any generalized point $p$ on a generalized circle $K$ and any generalized point $q \notin K$ there exists exactly one generalized circle $L$ through $q$ with $L \cap K = \{p\}$.

**M3.** There are four generalized points not on a generalized circle, and each generalized circle contains at least three generalized points.

The axioms **M1**, **M2**, and **M3** are called *incidence axioms*, and the sets $\mathfrak{P}$ and $\mathfrak{C}$ the *point set* and the *circle set* of $(\mathfrak{P}, \mathfrak{C})$. Usually a Möbius plane is identified with its point set. Two generalized circles *touch* each other if they have exactly one common generalized point. If they have two common generalized points, they *intersect properly*. If a Möbius plane, whose point and circle sets carry $T_1$-topologies, satisfies additional four axioms, called *continuity axioms*, then it is said to be a *topological Möbius plane*. Roughly speaking this means that the functions of joining, intersecting, and touching, as obtainable from M1 and M2, are continuous, the domain of proper intersecting is open in $\mathfrak{C} \times \mathfrak{C}$, and touching is the limit case of proper intersecting. For the precise formulations of the continuity axioms we refer to [101]. For our consideration it is essential that if the point set of a Möbius plane $(\mathfrak{P}, \mathfrak{C})$ is homeomorphic to the 2-sphere $S^2$ and each circle $K \in \mathfrak{C}$ is homeomorphic to the 1-sphere $S^1$, then $(\mathfrak{P}, \mathfrak{C})$ is a topological Möbius plane; see [101, Korollar 7.6]. A topological Möbius plane whose point set is a 2-manifold (i.e., locally homeomorphic to $\mathbb{R}^2$) is called a *flat Möbius plane*.

Let now $(\mathbb{M}^2, \|\cdot\|)$ be a strictly convex, smooth normed plane. Let a formal point at infinity $\infty$ be added to the plane $\mathbb{M}^2$, and let all lines of $\mathbb{M}^2$ pass through $\infty$. If $\mathfrak{P} = \mathbb{M}^2 \cup \{\infty\}$

## 3.1. KISSING SPHERES

and $\mathfrak{C}$ be the set of all circles and lines of $(\mathbb{M}^2, \|\cdot\|)$, we will prove that the incidence structure $(\mathfrak{P}, \mathfrak{C})$ is a flat Möbius plane. In the sequence we mean, saying generalized points of $(\mathbb{M}^2, \|\cdot\|)$ usual points or $\infty$. The circles and the lines of $(\mathbb{M}^2, \|\cdot\|)$ are referred to as generalized circles of $(\mathbb{M}^2, \|\cdot\|)$. Note that two intersecting lines, treated as elements of $\mathfrak{C}$, intersect properly, and that two parallel lines touch each other. For our purpose we need the next lemma.

**Lemma 3.1.1.** *In a strictly convex, smooth normed plane $(\mathbb{M}^2, \|\cdot\|)$ let there be given a segment $[p,q]$ and a line $G$ through $p$. If there exists a line $K$ through $p$ such that $K \neq \mathrm{L}(p,q)$ and $G \dashv K$, then the bisector $B(p,q)$ of $p$ and $q$ intersects the line $G$ in exactly one point.*

*Proof.* First note that anticircles of a strictly convex, smooth Minkowski plane are strictly convex, smooth curves. Theorem 8 in [71] states that the bisector $B(p,q)$ is contained in the interior of the strip with bounding lines $H_1$ and $H_2$, which are tangent to the anticircle

$$\mathcal{C}_a\left(\frac{p+q}{2}, \frac{\|p-q\|_a}{2}\right)$$

at the points $p$ and $q$. Thus we have $\mathrm{L}(p,q) \dashv_a H_1, H_2$, and by (1.5) it follows that $H_1, H_2 \dashv \mathrm{L}(p,q)$. Assume that $H_1 \equiv G$. Since $G \dashv K$, the smoothness of $(\mathbb{M}^2, \|\cdot\|)$ implies that $\mathrm{L}(p,q) \equiv K$, a contradiction. Therefore $H_1$ does not coincide with $G$, and $H_2$ has to intersect $G$. Let $H_2 \cap G = \{p_2\}$. Then the interior of conv $\{p, q, p_2\}$ contains a part of $B(p,q)$. But $B(p,q)$ is an unbounded curve, and it has to intersect $[p, p_2]$ in a point $x$. Assume that there exists a point $y \in B(p,q) \cap G$ and $y \neq x$. Let $y \in \mathrm{R}_p^+(x)$, say. Then

$$\|x-y\| = \|x-p\| - \|y-p\|.$$

Applying Lemma 1.2.6 for the circles $\mathcal{C}(x, \|x-p\|)$ and $\mathcal{C}(y, \|y-p\|)$, we get that they touch each other at $p$. But $q \in \mathcal{C}(x, \|x-p\|) \cap \mathcal{C}(y, \|y-p\|)$, which means that the point $x$ is unique. □

**Theorem 3.1.1.** *Each strictly convex, smooth normed plane $(\mathbb{M}^2, \|\cdot\|)$ is a Möbius plane.*

*Proof.* Let $\mathfrak{P}$ and $\mathfrak{C}$ be defined as above. Axiom **M3** is trivial, and Theorem 1.2.1 implies that **M1** is satisfied. Thus it only remains to check **M2**. Let $K \in \mathfrak{C}$, $p \in K$, and $q \notin K$. We distinguish two cases:

(i) $K$ is a line. Clearly, if $p \not\equiv \infty$ and $\mathrm{L}(p,q) \dashv K$, then $\mathcal{C}(\frac{p+q}{2}, \frac{\|p-q\|}{2})$ is the unique circle through $q$ touching $K$ at $p$. If the line $\mathrm{L}(p,q)$ is not orthogonal to $K$, consider the line $G$ through $p$ being orthogonal to $K$. Then, by Lemma 3.1.1, the bisector $B(p,q)$ intersects $G$ in exactly one point $x$. Therefore we have exactly one circle $\mathcal{C}(x, \|x-p\|)$ through $q$ and touching $K$ at $p$. If $p \equiv \infty$, then there exists also a unique element of $\mathfrak{C}$ through $\infty$ and $q \notin K$ having no further common points with $K$, namely the line through $q$ and parallel to $K$.

(ii) $K$ is a Minkowski circle with center $x$. If $q \equiv \infty$, the tangent of $K$ at $p$ is the unique element of $\mathfrak{C}$ which passes through $p, q$ and touches $K$. If $q \not\equiv \infty$, then the case $q \in \mathrm{L}(p,x)$

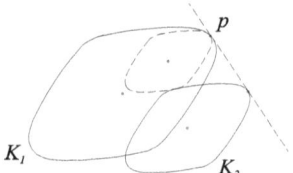

Figure 3.1: Theorem 3.1.3, (1). One of the generalized circles passing through $p$ and touching $K_1$, $K_2$ is a line.

is trivial. If $q \notin (p,x)$, consider the tangent $T$ of $K$ at $p$. Again by Lemma 3.1.1 there exists a unique circle $\mathcal{C}(y, \|y - p\|)$ through $p$ and $q$ which touches $T$. Moreover, the center $y$ of this circle lies on the line $\mathrm{L}(x,p)$, and by Lemma 1.2.6 we have that $K$ and $\mathcal{C}(y, \|y-p\|)$ touch each other. Thus we have proved that $(\mathfrak{P}, \mathfrak{C})$ is a Möbius plane. □

**Corollary 3.1.1.** *Each strictly convex, smooth Minkowski plane* $(\mathbb{M}^2, \| \cdot \|)$ *is a flat Möbius plane.*

*Proof.* Since $\mathbb{M}^2 \cup \infty$ is homeomorphic to the sphere $S^2$ (see, e.g., [12, p. 92, Proposition 4.3.6]), every line and circle in $(\mathbb{M}^2, \| \cdot \|)$ is homeomorphic to $S^1$ (see also [12, p. 92, Proposition 4.3.6 and p. 343, Corollary 11.3.4]). Therefore, by [101, Korollar 7.6], $(\mathfrak{P}, \mathfrak{C})$ is a topological Möbius plane which is evidently a flat Möbius plane. □

**Remark 3.1.1.** *Let $G$ be a line of a strictly convex, smooth normed plane and $p$ be a point on $G$. Then, by Theorem 3.1.1, for any point $q \notin G$ there exists a circle passing through $q$ and touching $G$ at $p$. Thus we obtain that the set of all generalized circles through two different points covers the plane.*

**Remark 3.1.2.** *A Möbius plane whose point set is homeomorphic to $S^2$ and whose circles are homeomorphic to $S^1$ is called a* spherical Möbius plane. *Clearly, every strictly convex, smooth normed plane is also a spherical Möbius plane. For spherical Möbius planes we refer to [97] and [98].*

### 3.1.2 Spheres kissing three given spheres

The next theorems, due to Groh [40], refer to flat Möbius planes. Theorem 3.1.2 below has been proved also by Strambach [98], for the case of spherical Möbius planes. In view of our Theorem 3.1.1 and Remark 3.1.2 we rewrite Groh's and Strambach's results for strictly convex, smooth normed planes.

## 3.1. KISSING SPHERES

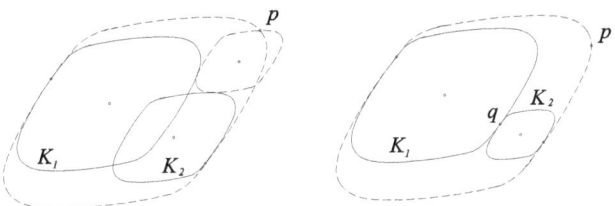

Figure 3.2: On the left Theorem 3.1.3, (1), and on the right Theorem 3.1.3, (2.1).

**Theorem 3.1.2.** *Let $K$ be a generalized circle in a strictly convex, smooth normed plane, and $p_1, p_2 \in \mathbb{M}^2 \setminus K$ be two points not separated by $K$. Then there exist exactly two generalized circles through $p_1$ and $p_2$ kissing $K$.*

**Theorem 3.1.3.** *Let $K_1 \neq K_2$ be two generalized circles in a strictly convex, smooth normed plane $(\mathbb{M}^2, \|\cdot\|)$. Assume that the point $p \in \mathbb{M}^2$ does not belong to $K_1 \cap K_2$, and $V_p$ be the connected component of $\mathbb{M}^2 \setminus (K_1 \cup K_2)$ containing $p$. Then there exist exactly $n$ generalized circles through $p$ kissing $K_1$ and $K_2$, where*

*(1) $n = 2$, if $K_1$ and $K_2$ intersect properly; see Figure 3.1 and Figure 3.2 (left);*

*(2) if $K_1$ and $K_2$ touch in $q$*

    *(2.1) and $p \in K_1 \cup K_2$, then $n = 1$; see Figure 3.2 (right);*

    *(2.2) and $p \notin K_1 \cup K_2$ and*

        *(2.2.1) for $i = 1, 2$, $K_i \cap \operatorname{bd} V_p \neq \{q\}$, then $n = 3$;*

        *(2.2.2) for some $i = 1, 2$, $K_i \cap \operatorname{bd} V_p = \{q\}$, then $n = 1$;*

*(3) if $K_1$ and $K_2$ do not intersect*

    *(3.1) and $p \in K_1 \cup K_2$, then $n = 2$;*

    *(3.2) and $p \notin K_1 \cup K_2$ and*

        *(3.2.1) for $i = 1, 2$, $K_i \cap \operatorname{bd} V_p \neq \emptyset$, then $n = 4$;*

        *(3.2.2) for some $i = 1, 2$, $K_i \cap \operatorname{bd} V_p = \emptyset$, then $n = 0$.*

Let $K_1, K_2, K_3$ be three pairwise intersecting generalized circles (i.e., properly intersecting or touching) in $(\mathbb{M}^2, \|\cdot\|)$. A set $T \subset \mathbb{M}^2$ is called a *circular triangle* if it is a connected component of $\mathbb{M}^2 \setminus (\cup_{i=1}^{3} K_i)$ such that each $K_i \cap \operatorname{bd} T$ is connected and has nonempty interior in $K_i$; see Figure 3.3.

**Theorem 3.1.4.** *In a strictly convex, smooth Minkowski plane, let there be given a circular triangle $T$ determined by the generalized circles $K_1, K_2$, and $K_3$. Then there exists precisely one generalized circle $K$ kissing $K_1, K_2$, and $K_3$, which belong to the closure of $T$.*

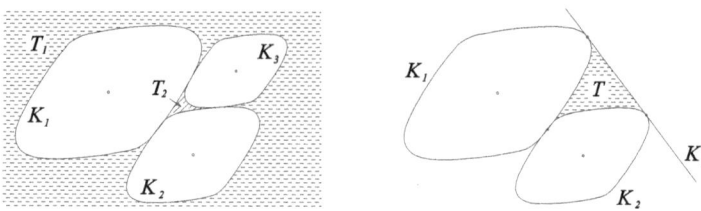

Figure 3.3: $T_1$, $T_2$, and $T$ are circular triangles.

Now we consider a strictly convex normed plane $(\mathbb{M}^2, \|\cdot\|)$, without the restriction of smoothness, and a 3-member kissing family $\mathfrak{F}$ of circles and lines. We are looking for a circle $C$ such that $\mathfrak{F} \cup C$ forms also a kissing family when the members of $\mathfrak{F}$ are in special position to each other. First we need some preliminaries. As it was announced in Theorem 1.2.1 not for any three non-collinear points there exists a circle containing them. But the next lemma shows that if the given points form an equilateral triangle, then this is true.

**Lemma 3.1.2.** *An equilateral triangle in a strictly convex Minkowski plane $(\mathbb{M}^2, \|\cdot\|)$ possesses exactly one circumcircle.*

*Proof.* Let there be given a triangle $\mathcal{T}(p_1, p_2, p_3)$ which is equilateral in $(\mathbb{M}^2, \|\cdot\|)$, and $m_1$ and $m_2$ be the midpoints of $[p_2, p_3]$ and $[p_3, p_1]$, respectively. Clearly, $p_1 \in B(p_2, p_3)$ and $p_2 \in B(p_3, p_1)$. The bisector $B(p_2, p_3)$ of $p_2$, $p_3$ is contained in the double cone of $p_2$ and $p_3$ with apex $p_1$, and $B(p_3, p_1)$ is contained in the double cone of $p_3$ and $p_1$ with apex $p_2$; see [46] and [72, Proposition 17]. The curve $B(p_2, p_3)$ can intersect the segment $[p_2, p_3]$ only in $m_1$. This means that if we denote the part of $B(p_2, p_3)$ between $p_1$ and $m_1$ by $\gamma_1$, then $\gamma_1 \subset \text{conv}\{p_1, p_2, p_3\}$. Analogously, if $\gamma_2$ is the part of $B(p_3, p_1)$ between $p_2$ and $m_2$, then $\gamma_2 \subset \text{conv}\{p_1, p_2, p_3\}$. If
$$J := [p_1, p_2] \cup [p_2, m_1] \cup \gamma_1,$$
then $J$ is a Jordan curve, with respect to which the point $m_2$ is not an interior point. Assuming that $m_2 \in \gamma_1 \subset B(p_2, p_3)$, we have
$$\|p_2 - m_2\| = \|m_2 - p_3\| \iff \|p_2 - \tfrac{1}{2}(p_1 + p_3)\| = \tfrac{1}{2}\|p_1 - p_3\| \iff \|p_2 - p_1 + p_2 - p_3\| = \|p_1 - p_3\|.$$
This is impossible, since the normed plane under consideration is strictly convex. Hence $m_2$ is an exterior point with respect to $J$. On the other hand, there exists an $\varepsilon > 0$ such that $\mathcal{C}(p_2, \varepsilon) \cap \gamma_2 \neq \emptyset$. If $q \in \mathcal{C}(p_2, \varepsilon) \cap \gamma_2$, then $q$ lies in the interior of $J$. This means that the part of $\gamma_2$ between $q$ and $m_2$ has to intersect $J$. But it does not intersect $[p_1, p_2]$ or $[p_2, m_1]$ (eventually it can touch $[p_1, p_2]$ or $[p_2, m_1]$); therefore it intersects $\gamma_1$. Thus we have shown that $B(p_2, p_3)$ and $B(p_3, p_1)$ have a common point, which completes the proof. □

The proof of Lemma 3.1.2 implies

**Lemma 3.1.3.** *The circumcenter of any equilateral triangle in a strictly convex normed plane lies the the interior of this triangle.*

**Theorem 3.1.5.** *In a strictly convex normed plane, let there be given three circles $\mathcal{C}(p_i, \lambda)$, $i = 1, 2, 3$, forming a kissing family. Then there exist exactly two circles which kiss the given circles.*

*Proof.* By Lemma 1.2.6 we have that the points $p_1, p_2, p_3$ form an equilateral triangle of side length $2\lambda$. Lemma 3.1.2 guarantees the existence of the circumcircle of the triangle $\mathcal{T}(p_1, p_2, p_3)$. Denote it by $\mathcal{C}(x, \mu)$. Moreover, Lemma 3.1.3 implies that $x \in \text{int} (\text{conv } \{p_1, p_2, p_3\})$. Thus we get

$$2\mu = \|p_1 - x\| + \|x - p_2\| > \|p_1 - p_2\| = 2\lambda,$$

and Lemma 1.2.6 yields that the circle $\mathcal{C}(x, \mu - \lambda)$, as well as $\mathcal{C}(x, \mu + \lambda)$, touches $\mathcal{C}(p_i, \lambda)$, $i = 1, 2, 3$. Assume that there exists a circle $\mathcal{C}(y, \nu)$ with $y \neq x$ which touches $\mathcal{C}(p_i, \lambda)$, $i = 1, 2, 3$. Clearly, $\mathcal{C}(y, \nu)$ and any of the circles $\mathcal{C}(p_i, \lambda)$, $i = 1, 2, 3$, cannot internally touch each other. Therefore $\mathcal{C}(y, \nu)$ externally touches the three circles $\mathcal{C}(p_i, \lambda)$, $i = 1, 2, 3$. But this contradicts the fact that in a strictly convex normed plane a triangle possesses at most one circumcenter; see Theorem 1.2.1. □

**Lemma 3.1.4.** *In a strictly convex normed plane $(\mathbb{M}^2, \|\cdot\|)$, let there be given two distinct points $p_1$, $p_2$ and two positive real numbers $\lambda_1$, $\lambda_2$ such that $\lambda_1 + \lambda_2 = \|p_1 - p_2\|$. Then the set*

$$\{x \in \mathbb{M}^2 : \|p_1 - x\| = \lambda_1 + \chi, \|p_2 - x\| = \lambda_2 + \chi, \text{ where } \chi \in \mathbb{R}^+\}$$

*is an unbounded curve.*

*Proof.* For any $\chi > 0$ the circles $\mathcal{C}(p_1, \lambda_1 + \chi)$ and $\mathcal{C}(p_2, \lambda_2 + \chi)$ have exactly two common points which lie in the different half-planes with respect $L(p_1, p_2)$. Denote these points by $x_1$ and $x_2$, and let $H^+$ and $H^-$ be both half-planes bounded by $L(p_1, p_2)$. Assume that $x_1 \in H^+$ and $x_1 \in H^+$. Thus we have the mapping $\varphi : \mathbb{R}^+ \longrightarrow H^+$, which assigns to every real positive number $\chi$ the point of $\mathcal{C}(p_1, \lambda_1 + \chi) \cap \mathcal{C}(p_2, \lambda_2 + \chi)$ which lies in $H^+$. It is easy to see that $\varphi$ is continuous in both directions. Indeed, for any convergent sequence $\{\chi_n\}_{n=1}^\infty \longrightarrow \chi$ we have

$$\{\lambda_1 + \chi_n\}_{n=1}^\infty \longrightarrow \lambda_1 + \chi \iff \{\|p_1 - \varphi(\chi_n)\|\}_{n=1}^\infty \longrightarrow \|p_1 - \varphi(\chi)\| \iff$$

$$\{p_1 - \varphi(\chi_n)\}_{n=1}^\infty \longrightarrow p_1 - \varphi(\chi) \iff \{\varphi(\chi_n)\}_{n=1}^\infty \longrightarrow \varphi(\chi).$$

Thus we obtain that $\varphi([0, +\infty))$ is a curve. It is also evident that this curve is unbounded. □

**Theorem 3.1.6.** *In a strictly convex Minkowski plane* $(\mathbb{M}^2, \|\cdot\|)$, *let there be given two circles* $\mathcal{C}(p_1, \lambda_1)$, $\mathcal{C}(p_2, \lambda_2)$ *touching each other externally. If* $G$ *is their common supporting line, then there exists a circle* $\mathcal{C}(x, \lambda)$ *kissing* $\mathcal{C}(p_1, \lambda_1)$, $\mathcal{C}(p_2, \lambda_2)$, *and* $G$.

*Proof.* Denote by $y$ the touching point of $\mathcal{C}(p_1, \lambda_1)$ and $\mathcal{C}(p_2, \lambda_2)$. Then $y$ lies on the segment $[p_1, p_2]$; see Lemma 1.2.6. If $q_i$, $i = 1, 2$, is the touching point of $\mathcal{C}(p_i, \lambda_i)$ and $G$, then $L(p_i, q_i) \dashv G$. Consider the part $B^+(\lambda_1, \lambda_2)$ of the curve

$$B(\lambda_1, \lambda_2) = \{x \in \mathbb{M}^2 : \|p_1 - x\| = \lambda_1 + \chi, \|p_2 - x\| = \lambda_2 + \chi, \text{ where } \chi \geq 0\},$$

which lies in the half-plane with respect to $L(p_1, p_2)$ containing the points $q_1$ and $q_2$. Clearly, $B^+(\lambda_1, \lambda_2)$ passes through $y$ and does not intersect the segments $[p_1, q_1]$ and $[p_2, q_2]$. Thus $B^+(\lambda_1, \lambda_2)$ has to intersect the segment $[q_1, q_2]$. Denote the first intersection point of $B^+(\lambda_1, \lambda_2)$ and $[q_1, q_2]$ by $z$. Let $z$ be attained for $\chi = \chi_0$, i.e., $z = \varphi(\chi_0)$, where $\varphi$ is defined as in Lemma 3.1.4. Consider the function $f : [0, \chi_0] \longrightarrow \mathbb{R}$ defined by $f(\chi) := d(\varphi(\chi), G)$. If $f$ is continuous, since $f(0) = d(y, G) > 0$, $f(\chi_0) = 0$, then there exists a number $\chi^* \in [0, \chi_0]$ such that $f(\chi^*) = \chi^*$. Thus for $x^* = \varphi(\chi^*)$ we have

$$\|p_1 - x^*\| = \lambda_1 + \chi^*, \|p_2 - x^*\| = \lambda_2 + \chi^*, \text{ and } d(x^*, G) = \chi^*.$$

This means that the circle $\mathcal{C}(x^*, \chi^*)$ kisses $\mathcal{C}(p_1, \lambda_1)$, $\mathcal{C}(p_2 \lambda_2)$, and $G$. In order to complete the proof we need to show that the function $f$ is continuous. In Lemma 3.1.4 we have proved that for the convergent sequence $\{\chi_n\}_{n=1}^\infty \longrightarrow \chi$ the sequence $\{x_n = \varphi(\chi_n)\}_{n=1}^\infty$ is also convergent, and that it converges to $x = \varphi(\chi)$. Denote by $y_n$ and $y$ those points of $G$ such that $L(x_n, y_n) \dashv G$ and $L(x, y) \dashv G$. Then $f(\chi_n) = \|x_n - y_n\|$ and $f(\chi) = \|x - y\|$. Let $\varepsilon \in \mathbb{R}^+$. For any $n \in \mathbb{N}$ there exists an $m \in \mathbb{N}$ with $m > n$ for which $x_m \in \text{int } C(x, \varepsilon)$. Since in a strictly convex normed plane the common supporting line of $\mathcal{C}(x, \varepsilon)$ and $\mathcal{C}(y, \varepsilon)$ is parallel to the line $L(x, y)$, then $y_m \in \mathcal{D}(y, \varepsilon)$. This means that $\{y_n\}_{n=1}^\infty \longrightarrow y$, which implies $\{f(\chi_n)\}_{n=1}^\infty \longrightarrow f(\chi)$. □

## 3.2 Covering a disc by translates of the unit disc

Let $K$ be a convex body, and by $h_k(K)$ denote the smallest positive ratio of $k$ homothetical copies of $K$ whose union covers $K$. For $k \in \{3, 4\}$ the following bounds on $h_k(K)$ are known:

$$\frac{2}{3} \leq h_3(K) \leq 1, \tag{3.1}$$

$$\frac{1}{2} \leq h_4(K) \leq \frac{\sqrt{2}}{2}; \tag{3.2}$$

## 3.2. Covering a disc by translates of the unit disc

see [54]. In this section we give an exact geometric description of $h_k(K)$ for $k = \{3, 4\}$, with $K$ being a strictly convex, centrally symmetric convex body in terms of the radius of inscribed equilateral polygon. For that reason we reformulate the above problem for centrally symmetric convex bodies $K$, i.e., $K$ can be considered as unit disc $\mathcal{D}$ with respect to some norm. Let $R_k(\mathcal{D})$ be the maximal radius of all homothets of $\mathcal{D}$ that can be covered by $k$ translates of $\mathcal{D}$. Then

$$R_k(\mathcal{D}) = \frac{1}{h_k(\mathcal{D})}, \qquad (3.3)$$

and one can rewrite the inequalities (3.1), (3.2) as follows:

$$1 \leq R_3(\mathcal{D}) \leq \frac{3}{2}, \quad \sqrt{2} \leq R_4(\mathcal{D}) \leq 2. \qquad (3.4)$$

### 3.2.1 Lemmas

**Lemma 3.2.1.** *Let there be given a convex body $K$ in a normed plane $(\mathbb{M}^2, \|\cdot\|)$, and $\mathfrak{K} = \{K_i\}_{i=1}^k$ be a covering of $K$. If $x \in \operatorname{bd} K_i \cap K$, where $i \in \{1, \ldots, k\}$, then there exists a body $K_j$ from $\mathfrak{K}$ different to $K_i$ such that $x \in K_j$.*

*Proof.* We argue by contradiction. Suppose that for any $j = 1, \ldots, k$ and $j \neq i$ we have $x \notin K_j$. Then there exists a disc $\mathcal{D}(x, \varepsilon)$ with $(\operatorname{int} \mathcal{D}(x, \varepsilon)) \cap B_j = \emptyset$. Let $y$ be a point of int $\mathcal{D}(x, \varepsilon)$ such that $y \notin K_i$. Denote by $D$ a disc centered at $y$ whose interior lies in int $\mathcal{D}(x, \varepsilon)$ and which also satisfies $D \cap K_i = \emptyset$. For any $j = 1, \ldots, k$, and $j \neq i$, there exists a point $y_j \in K_j \cap B$ with $\|y - y_j\| = \inf\{\|y - z\| : z \in K_j \cap K\}$; see, e.g., [100, p. 45, Theorem 1.9.1]. If $y_0$ is that point among $\{y_j\}_{j=1, j \neq i}^k$ which has the smallest distance to $y$, then clearly $y_0 \notin \operatorname{int} \mathcal{D}(x, \varepsilon)$. Thus, for any

$$z \in \bigcup_{j \neq i} (K_j \cap K)$$

we get $\|y - y_0\| \leq \|y - z\|$. Let $y^*$ be a point lying in $[y, y_0] \cap D$. Then $\|y - y^*\| < \|y - y_0\|$, which means that

$$y^* \notin \bigcup_{j \neq i} (K_j \cap K).$$

Since $y^* \in D$, we get $y^* \notin K_i$. Besides this, the convexity of $D$ implies that the point $y^*$ belongs to $K$. This contradicts the fact that $\{K_i\}_{i=1}^k$ is a covering of $K$. □

**Lemma 3.2.2.** *Let $p$ be a point in a strictly convex normed plane $(\mathbb{M}^2, \|\cdot\|)$ different to the origin $0$. Then the bisector of the points $p$ and $-p$ intersects the circle $\mathcal{C}(0, \|p\|)$ in exactly two points.*

*Proof.* It is clear that $B(-p,p)$ intersects $\mathcal{C}(0,\|p\|)$ in at least two points which are opposite points of that circle, e.g., in $x$ and $-x$. We will show that, besides $x$ and $-x$, there are no further intersection points of $B(-p,p)$ and $\mathcal{C}(0,\|p\|)$. We have that the bisector $B(-p,p)$ is contained in the double cone $V$ of $p$ and $-p$ with apex $x$; see [46] and [72, Proposition 17]. Let us consider that part of $V$ (denoted by $V^*$) which lies on the half-plane $\mathrm{HP}_x^+(-p,p)$. Thus

$$V^* = \{x + \lambda(-p-x) + \mu(p-x) : \lambda, \mu \leq 0\} \cup \{x + \lambda(-p-x) + \mu(p-x) : \lambda, \mu \in (0,1)\}.$$

We will show that neither

$$V^- = \{x \in \mathbb{M}^2 : x + \lambda(-p-x) + \mu(p-x) : \lambda, \mu \leq 0\}$$

nor the set

$$V^+ = \{x \in \mathbb{M}^2 : x + \lambda(-p-x) + \mu(p-x) : \lambda, \mu \in (0,1)\}$$

contains points of $\mathcal{C}(0,\|p\|)$ which are different to $x$. For any $y \in V^-$ we have

$$y = x + \lambda(-p-x) + \mu(p-x) \iff (1-\lambda-\mu)\,x = y + (\lambda-\mu)\,p,$$

where $\lambda, \mu \leq 0$. Therefore

$$(1-\lambda-\mu)\,\|x\| = \|y + (\lambda-\mu)\,p\| < \|y\| + |\lambda-\mu|\,\|p\| \iff$$
$$\begin{cases} \|x\| \leq (1-2\lambda)\,\|x\| < \|y\| & \text{if } \lambda-\mu \geq 0, \\ \|x\| \leq (1-2\mu)\,\|x\| < \|y\| & \text{if } \lambda-\mu < 0. \end{cases}$$

Hence any $y \in V^-$ is an exterior point with respect to $\mathcal{C}(0,\|p\|)$. On the other hand, $V^+$ is a triangle inscribed to the strictly convex curve $\mathcal{C}(0,\|p\|)$. Therefore $V^+$ also does not contain points of $\mathcal{C}(0,\|p\|)$ (except for $p$, $-p$, and $x$). $\square$

**Lemma 3.2.3.** *In a strictly convex normed plane let there be given two points $x_1$ and $x_3$ which are opposite points of the unit circle $\mathcal{C}$. If*

$$B(x_1,x_3) \cap \mathcal{C} = \{x_2, x_4\},$$

*then the intersection points of $\mathcal{C}(x_i,1)$ and $\mathcal{C}(x_{i+1},1)$ (with $i = 1,2,3,4$ and $x_5 \equiv x_1$), which are different to $0$, lie on the same circle $C$ with radius between $1$ and $2$.*

*Proof.* Note that according to Lemma 3.2.2 the intersection of $B(x_1,x_3)$ and $\mathcal{C}$ consists of exactly two points, which are opposite in $\mathcal{C}$. The monotonicity lemma implies $\|x_i - x_{i+1}\| < 2$, $i = 1,\ldots,4$. Hence $\mathcal{C}(x_i,1)$ and $\mathcal{C}(x_{i+1},1)$ have exactly two points in common. Clearly, the

## 3.2. COVERING A DISC BY TRANSLATES OF THE UNIT DISC

origin 0 is one of them, and we denote by $p_i$ the other intersection point. Thus, by Lemma 1.2.7 in Chapter 1 we obtain $x_i + x_{i+1} = p_i$. Since $x_1$ and $x_3$ are opposite points, like also $x_2$ and $x_4$, it follows that $\|p_i\| = \|x_{i+1} - x_{i+2}\|$ (note that $x_6 \equiv x_2$). But we have $x_2, x_4 \in B(x_1, x_3)$, i.e., $\|x_1 - x_2\| = \|x_2 - x_3\| = \|x_3 - x_4\| = \|x_4 - x_1\| = \lambda$, which is equivalent to $p_i \in \mathcal{C}(0, \lambda)$. Moreover, with respect to the triangle $\mathcal{T}(x_1, x_2, x_3)$ we have $2 = \|x_1 - x_3\| < \|x_1 - x_2\| + \|x_2 - x_3\| \iff 1 < \lambda$. On the other hand, applying Lemma 1.2.4 for the convex quadrangle $\mathcal{Q}(x_1, x_2, x_3, x_4)$, we obtain

$$\|x_1 - x_3\| + \|x_2 - x_4\| > \|x_1 - x_2\| + \|x_4 - x_3\| \iff 2 > \lambda.$$

□

**Lemma 3.2.4.** *In a normed plane* $(\mathbb{M}^2, \|\cdot\|)$, *let there be given two circles* $\mathcal{C}(x_1, \lambda_1)$ *and* $\mathcal{C}(x_2, \lambda_2)$ *with* $\lambda_1 \neq \lambda_2$. *Then the homothety*

$$\varphi : x \longmapsto \frac{-\lambda_2 x_1 + \lambda_1 x_2}{\lambda_1} + \frac{\lambda_2}{\lambda_1} x$$

*maps* $\mathcal{C}(x_1, \lambda_1)$ *into* $\mathcal{C}(x_2, \lambda_2)$. *The center of* $\varphi$ *is the point*

$$s = \frac{\lambda_2}{\lambda_2 - \lambda_1} x_1 - \frac{\lambda_1}{\lambda_2 - \lambda_1} x_2.$$

The proof of this lemma is immediate.

### 3.2.2 Results

**Theorem 3.2.1.** *In a strictly convex normed plane* $(\mathbb{M}^2, \|\cdot\|)$ *let there be given an equilateral triangle* $\mathcal{T}(p_1, p_2, p_3)$ *of side-length 2. Then the circumradius of* $\mathcal{T}(p_1, p_2, p_3)$ *is* $> 1$, *and the circumdisc of* $\mathcal{T}(p_1, p_2, p_3)$ *can be covered by three translates of the unit discs.*

*Proof.* If $\mathcal{C}(q, \lambda)$ is the circumcircle of $\mathcal{T}(p_1, p_2, p_3)$ (note that according to Lemma 3.1.2 this circumcircle exists), then

$$\|p_1 - q\| + \|q - p_2\| > \|p_1 - p_2\| \iff 2\lambda > 2 \iff \lambda > 1.$$

We now show that if $m_i$ is the midpoint of $[p_j, p_k]$, where $\{i, j, k\} = \{1, 2, 3\}$, then the discs $\mathcal{D}(m_i, 1)$, $i = 1, 2, 3$, cover $\mathcal{D}(q, \lambda)$. At first we check whether $m_j, m_k \in \mathcal{C}(m_i, 1)$. Indeed,

$$\|m_i - m_j\| = \left\|\frac{p_j + p_k}{2} - \frac{p_k + p_i}{2}\right\| = \frac{1}{2}\|p_j - p_i\| = 1.$$

Thus we get that the discs $\mathcal{D}(m_i, 1)$, $i = 1, 2, 3$, cover conv $\{p_1, p_2, p_3\}$. In order to complete the proof, it remains to show that $\mathcal{D}(m_3, 1)$ covers the circular arc $\text{arc}(p_1, p_2; q)$, say. Let us assume that $q \equiv 0$ and write $C_1 := \mathcal{C}(0, \lambda)$, $C_2 := C(m_3, 1)$. We consider the homothety $\varphi$ mapping the circle $C_1$ into the circle $C_2$, i.e.,

$$\varphi : x \longmapsto m_3 + \frac{1}{\lambda} x;$$

see Lemma 3.2.4. Then the center of $\varphi$ is $s = \frac{\lambda}{\lambda - 1} m_3$, which belongs to the ray of $\mathrm{R}_0^-(m_3)$, i.e., $s$ lies on the half plane $\mathrm{HS}_{p_3}^-(p_1, p_2)$. We will prove that $s \notin \mathcal{D}(m_3, 1)$ and $s \notin \mathcal{D}(0, \lambda)$. Assume that $s \in \mathcal{D}(m_3, 1)$. This is equivalent to

$$1 \geq \|m_3 - s\| = \|m_3 - \frac{\lambda}{\lambda - 1} m_3\| = \frac{1}{\lambda - 1} \|m_3\| = \frac{1}{2(\lambda - 1)} \|p_1 + p_2\| \iff$$

$$2(\lambda - 1) \geq \|p_1 + p_2\| \iff 2\lambda \geq \|p_1 + p_2\| + 2.$$

The last inequality contradicts the triangle inequality referring to $\mathcal{T}(p_1, p_2, -p_2)$. Assuming that $s \in \mathcal{D}(0, \lambda)$, we get

$$\|s\| = \frac{\lambda}{\lambda - 1} \|m_1\| = \frac{\lambda}{2(\lambda - 1)} \|p_1 + p_2\| \leq \lambda \iff 2(\lambda - 1) \geq \|p_1 + p_2\|,$$

again a contradiction.

Further on, let $x_1$ belong to the circular arc $\mathrm{arc}_\lambda(p_1, p_2; q)$ of $C_1$ not containing $p_3$. If $\mathrm{L}(s, x_1) \cap C_1 = \{x_1, y_1\}$, then $x_1$ is between $s$ and $y_1$. This follows from the fact that $s$ belongs to the same half-plane with respect to $\mathrm{L}(p_1, p_2)$ containing the circular arc $\mathrm{arc}_\lambda(p_1, p_2; q)$. If $\varphi(x_1) = x_2$, then $x_2 \in C_2$ and

$$x_2 = m_3 + \frac{1}{\lambda} x_1 \iff x_2 = \frac{\lambda - 1}{\lambda} s + \frac{1}{\lambda} x_1.$$

Since $\lambda > 1$, we state that the point $x_2$ is between $s$ and $x_1$. Let $\mathrm{L}(x_1, x_2) \cap C_2 = \{x_2, y_2\}$. This means that $x_2$ is between $s$ and $y_2$. Moreover, we have that $y_2 = \varphi(y_1)$, equivalent to the fact that $y_2$ is between $s$ and $y_1$. Thus we have that the points $s, x_2, x_1, y_2, y_1$ are located on the line $\mathrm{L}(x_1, x_2)$ in this order or in the order $s, x_2, y_2, x_1, y_1$. But the second situation is impossible, because the points $x_1$ and $y_2$ lie on different half-planes with respect to $\mathrm{L}(p_1, p_2)$. Therefore $x_1$ is between $x_2$ and $y_2$, equivalent to the fact that $\mathcal{D}(m_3, 1)$ covers the circular arc $\mathrm{arc}_\lambda(p_1, p_2; q)$. $\square$

The next theorem describes the geometric meaning of the maximal radius $R_3(\mathcal{D})$ of all homothetes of the unit disc $\mathcal{D}$ in a strictly convex normed plane that can be covered by 3 translates of $\mathcal{D}$.

**Theorem 3.2.2.** *If* $(\mathrm{M}^2, \|\cdot\|)$ *is a strictly convex normed plane with unit disc* $\mathcal{D}$*, then the quantity* $R_3(\mathcal{D})$ *is the maximal circumradius of equilateral triangles with side-length 2.*

## 3.2. COVERING A DISC BY TRANSLATES OF THE UNIT DISC

*Proof.* If $\lambda$ is the maximal circumradius of all equilateral triangles of side-length 2, we will prove that the disc $\mathcal{D}(0, \lambda + \varepsilon)$, where $\varepsilon > 0$, cannot be covered by three translates of the unit disc $\mathcal{D}$. Let $\mathcal{T}(p_1, p_2, p_3)$ be an equilateral triangle of side-length 2 inscribed in $\mathcal{D}(0, \lambda)$. If $\varphi$ is the homothety which maps $\mathcal{C}(0, \lambda)$ into $\mathcal{C}(0, \lambda + \varepsilon)$ with $\varphi(p_1, p_2, p_3) = p'_1, p'_2, p'_3$, then $\mathcal{T}(p'_1, p'_2, p'_3)$ is equilateral and of side-length $2 + \frac{2\varepsilon}{\lambda}$. Assume that $\mathcal{D}(0, \lambda + \varepsilon)$ can be covered by the translates $D_1$, $D_2$, and $D_3$ of $\mathcal{D}$. If $p'_1 \in D_1$, say, then $p'_2, p'_3 \notin D_1$, by the monotonicity lemma. If $p'_2 \in D_2$, say, then $p'_3 \notin D_2$. Therefore $p'_3 \in D_3$. For $\{i, j, k\} = \{1, 2, 3\}$, let arc$(p'_i, p'_j; 0)$ be a circular arc of $\mathcal{C}(0, \lambda + \varepsilon)$ between the points $p'_i$ and $p'_j$ which does not contain $p'_k$. Then the monotonicity lemma and Lemma 3.1.3 imply that any point of arc$(p'_i, p'_j; 0)$ does not belong to $D_k$. Let bd $D_1 \cap \mathcal{C}(0, \lambda + \varepsilon) = \{q_1, q_3\}$ and $q_1 \in$ arc$(p'_1, p'_2; 0)$, $q_3 \in$ arc$(p'_1, p'_3; 0)$. Note that if $q_1 \equiv q_3$, then $q_1 \equiv q_3 \equiv p_1$, which means that $\mathcal{D}(0, \lambda + \varepsilon)$ cannot be covered by $D_1, D_2$, and $D_3$. By Lemma 3.2.1 we have that $q_1 \in D_2$ and $q_3 \in D_3$. Thus we obtain $\mathcal{T}(q_1, q_2, q_3)$ inscribed in $C(0, \lambda + \varepsilon)$ such that $q_1 \in$ arc$(p'_1, p'_2; 0)$, $q_2 \in$ arc$(p'_2, p'_3; 0)$, $q_3 \in$ arc$(p'_3, p'_1; 0)$, $q_1 \neq p_1$, $q_1 \neq p_2$, $q_2 \neq p_2$, $q_2 \neq p_3$, $q_3 \neq p_3$, $q_3 \neq p_1$, and the sides of $\mathcal{T}(q_1, q_2, q_3)$ are of length $\leq 2$. Moreover, it is easy to see that the interior of conv $\{q_1, q_2, q_3\}$ contains the origin 0. Let $\mathcal{T}(q_1, q_2, q_3)$ be positively oriented, say. Construct an equilateral triangle $\mathcal{T}(u_1, u_2, u_3)$ of side-length 2, which is positively oriented and such that L$(u_1, u_2)$ is parallel to L$(q_1, q_2)$. According to [72, Proposition 33], for a given segment $[u_1, u_2]$ there exists exactly one such triangle. Let $\mathcal{C}(0, \mu)$ be a translate of the circumcircle of $\mathcal{T}(u_1, u_2, u_3)$. Then $\mu \leq \lambda$. If $v_1, v_2, v_3$ are the images of $u_1, u_2, u_3$ with respect this translation, let $v'_1, v'_2, v'_3$ be the images of $v_1, v_2, v_3$ with respect to the homothety mapping $\mathcal{C}(0, \mu)$ into $\mathcal{C}(0, \lambda + \varepsilon)$. Thus we obtain that $\mathcal{T}(v'_1, v'_2, v'_3)$ is an equilateral triangle inscribed in $\mathcal{C}(0, \lambda + \varepsilon)$ and having side-length $\frac{2(\lambda + \varepsilon)}{\mu} > 2$, where the side $[v'_1, v'_2]$ is parallel to $[q_1, q_2]$. Therefore the side $[v'_1, v'_2]$ lies in the open half-plane with respect to L$(q_1, q_2)$ which contains the origin 0. The third vertex $v'_3$ of $\mathcal{T}(v'_1, v'_2, v'_3)$ either belongs either to the circular arc arc$_{\lambda+\varepsilon}(q_1, q_3; 0)$, or to arc$_{\lambda+\varepsilon}(q_2, q_3; 0)$. But both these cases contradict the monotonicity lemma. $\square$

The next proposition gives an upper bound on $R_3(\mathcal{D})$, where $\mathcal{D}$ is the unit disc in a strictly convex normed plane. This upper bound strengthens the second inequality in (3.4) for the case that $B$ is centrally symmetric and strictly convex.

**Proposition 3.2.1.** *In a normed plane* $(\mathbb{M}^2, \|\cdot\|)$ *with unit disc* $\mathcal{D}$ *we have* $R_3(\mathcal{D}) \leq \frac{4}{3}$ *if* $\mathcal{D}$ *is strictly convex, and* $R_3(\mathcal{D}) = \frac{4}{3}$ *if* bd $\mathcal{D}$ *is an affine regular hexagon.*

*Proof.* Let $\pm p, \pm q, \pm (p+q)$ be the vertices of a hexagon which is regular in the norm (i.e., an affine regular hexagon with sides of the same Minkowskian length) and inscribed in the unit circle $\mathcal{C} = $ bd $\mathcal{D}$. Note that this is possible; see, e.g., the survey [72, § 4]. The triangle with vertices $\frac{4}{3} p - \frac{2}{3} q$, $-\frac{2}{3} p + \frac{4}{3} q$, and $-\frac{2}{3} (p+q)$ is equilateral with side-length 2 and inscribed in $\frac{4}{3}\mathcal{C}$. Therefore, if $\mathcal{C}$ is strictly convex, we have $R_3(\mathcal{D}) \leq \frac{4}{3}$, and $R_3(\mathcal{D}) = \frac{4}{3}$ holds if $\mathcal{C}$ is an affine regular hexagon. □

In case of four covering discs we shall see that, somehow analogous to the considerations up to now, regular quadrangles inscribed to a circle will play an essential role.

**Remark 3.2.1.** *It is easy to check that the points $p_1, p_2, p_3$, and $p_4$ (see the proof of Lemma 3.2.3) form a parallelogram all whose sides are of Minkowskian length 2 and whose two diagonals have the same length. The proof of Lemma 3.2.3 also implies that for any given direction such a parallelogram with two sides parallel to this direction can be constructed.*

**Theorem 3.2.3.** *If, in a strictly convex normed plane, $\mathcal{C}(x_i, 1)$, $i = 1, 2, 3, 4$, and $C$ are determined as in Lemma 3.2.3, then $\mathcal{C}(x_i, 1)$ with $i = 1, 2, 3, 4$ is a covering of $C$.*

*Proof.* In view of the constructions of $\mathcal{C}(x_i, 1)$, $i = 1, 2, 3, 4$, it is enough to prove that, e.g., $\mathcal{D}(x_1, 1)$ covers the circular arc of $C$ with endpoints $p_1$ and $p_4$. This can be verified quite similar as in the proof of Theorem 3.2.1, using the homothety that maps $\mathcal{C}(0, \lambda)$ into $\mathcal{C}(x_1, 1)$. □

Based on Remark 3.2.1, the next theorem can be proved in the same way as Theorem 3.2.2.

**Theorem 3.2.4.** *In a strictly convex normed plane with unit disc $\mathcal{D}$, the quantity $R_4(\mathcal{D})$ is the maximal circumradius of all parallelograms whose four sides are of Minkowskian length 2, and whose two diagonals have the same length.*

## 3.3 Regular 4-coverings

According to the famous conjecture of Hadwiger, completely confirmed only for the planar case, any convex body in the plane can be covered by 4 smaller positive homothets of itself. The smallest possible ratio of those four homothets is attained in case of the so-called *regular 4-covering*. This regular 4-covering was constructed by Lassak [53] in order to prove that the smallest possible ratio is $\frac{\sqrt{2}}{2}$. In this section we continue the investigations of Lassak on regular 4-coverings and derive further properties of them.

### 3.3.1 Properties of a regular 4-covering

We start with the construction of a regular 4-covering. For that purpose the notions of *quasi-dual* and *dual parallelograms* are needed. Two parallelograms $P$ and $Q$ in $\mathbb{M}^2$ are said to be *quasi-dual* if the sides of $P$ are parallel to the diagonals of $Q$ and the sides of $Q$ are parallel to the diagonals of $P$, i.e., the parallelograms $P$ and $Q$ are quasi-dual if and only if for some possible

## 3.3. REGULAR 4-COVERINGS

denotations of the vertices of $P$ and $Q$ (e.g., by $p_1, p_2, p_3, p_4$ and $q_1, q_2, q_3, q_4$, respectively) the relations

$$(p_1 - p_2) \parallel (q_2 - q_4), \quad (p_2 - p_3) \parallel (q_1 - q_3), \quad (q_1 - q_2) \parallel (p_1 - p_3), \quad (q_2 - q_3) \parallel (p_2 - p_4) \quad (3.5)$$

hold. The following lemma is proved in [53].

**Lemma 3.3.1.** *Let $P$ and $Q$ be parallelograms in $(\mathbb{M}^2, \|\cdot\|)$ with vertices $p_1, p_2, p_3, p_4$ and $q_1, q_2, q_3, q_4$, respectively. If they are quasi-dual, then*

(i) *any three conditions of (3.5) imply the fourth one;*

(ii) $\dfrac{\|p_1 - p_2\|}{\|q_2 - q_4\|} = \dfrac{\|p_2 - p_3\|}{\|q_3 - q_1\|} = \lambda, \quad \dfrac{\|q_1 - q_2\|}{\|p_1 - p_3\|} = \dfrac{\|q_2 - q_3\|}{\|p_2 - p_4\|} = \mu \quad \text{and} \quad \lambda\mu = \dfrac{1}{2}.$

Also the next statement was proved by Lassak.

**Lemma 3.3.2.** ([53, **Lemma 4**]) *Every convex body in the plane has a pair of inscribed quasi-dual parallelograms, where a diagonal of one of them can be of any given direction.*

Now we are ready to give a description of a regular 4-covering as it is constructed in [53].

Let $K$ be a convex body in the Euclidean plane. According to [53, Lemma 4] there exist points $p_1, q_1, \ldots, p_4, q_4$ lying in this order in bd $K$ such that $p_1, \ldots, p_4$ are the vertices of the parallelogram $P$, $q_1, \ldots, q_4$ are the vertices of the parallelogram $Q$, the parallelograms $P, Q$ are quasi-dual, and $\lambda \leq \frac{\sqrt{2}}{2}$, where $\lambda$ is determined by the first equation of (ii) in Lemma 3.3.1.

Let

$$L(q_4, p_1) \cap L(p_2, q_2) = \{t_1\}, \quad L(q_1, p_2) \cap (p_3, q_3) = \{t_2\},$$

$$L(q_2, p_3) \cap L(p_4, q_4) = \{t_3\}, \quad L(q_3, p_4) \cap L(p_1, q_1) = \{t_4\}; \quad (3.6)$$

see Figure 3.4.

Furthermore, let $\varphi_i$ with $i = 1, \ldots, 4$ be the homothety with center $t_i$ and ratio $\lambda$. Then $\cup_{i=1}^{4} \varphi_i(K)$ is a covering of $K$, called a *regular 4-covering* and denoted by $\text{cov}(P, Q, \lambda)$. Moreover, for $i = 1, \ldots, 4$ the covering $\varphi_i(K)$ contains the intersection point of the diagonals of $P$; see [53, § 3].

**Theorem 3.3.1.** *In the Euclidean plane, let there be given a convex body $K$ of diameter 1. If $\text{cov}(P, Q, \lambda)$ is a regular 4-covering of $K$, then the smallest homothetical copy of $K$, which contains $\text{cov}(P, Q, \lambda)$, is of diameter $2\lambda$.*

*Proof.* Let the vertices $p_1, \ldots, p_4$ and $q_1, \ldots, q_4$ of $P$ and $Q$, respectively, be placed on bd $K$ as shown in Figure 3.4. Let $t_i$, $i = 1, \ldots, 4$, be determined by (3.6), and $\varphi_i$ be the homothety with center $t_i$ and ratio $\lambda$. Then $\text{cov}(P, Q, \lambda) = \cup_{i=1}^{4} K_i$, where $K_i = \varphi_i(K)$, i.e.,

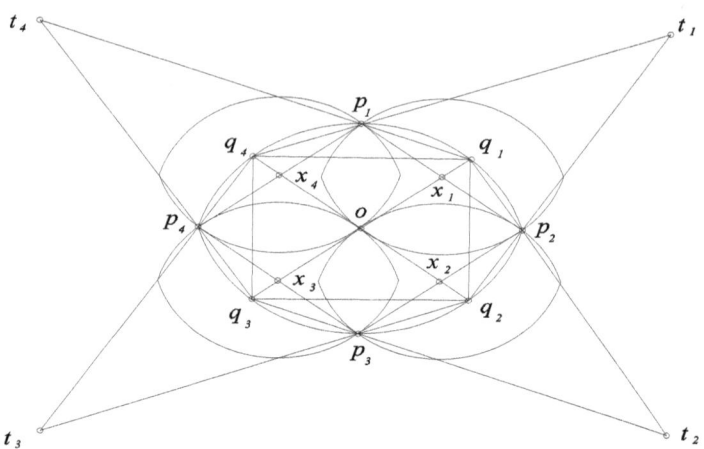

Figure 3.4: A regular 4-covering.

$$K_i = (1-\lambda)t_i + \lambda K. \tag{3.7}$$

Since $\varphi_i(\{q_{i-1}, q_{i+1}\}) = \{p_i, p_{i+1}\}$ (the subscripts of $p$ and $q$ are meant modulo 4), we have

$$p_i = (1-\lambda)t_i + \lambda q_{i-1}, \quad p_{i+1} = (1-\lambda)t_i + \lambda q_{i+1}. \tag{3.8}$$

The equations (3.7) and (3.8) yield

$$K_1 = p_2 - p_3 + K_3. \tag{3.9}$$

Consider the point $x = (1-\lambda)t_3 + \lambda q_1$ on the boundary of $K_3$. Thus by (3.8) we get

$$x - \frac{1}{2}(p_1 + p_3) = \frac{1}{2}(p_3 - p_1) - \lambda(q_2 - q_1) = \frac{1}{2}(p_3 - p_1) - \lambda\mu(p_3 - p_1),$$

where $\mu$ is determined in Lemma 3.3.1. Since $\lambda\mu = \frac{1}{2}$, we obtain that $x$ coincides with the intersection point of the diagonals of $P$. Therefore the point

$$q_1' = p_2 - p_3 + \frac{1}{2}(p_1 + p_3) = p_2 + \frac{1}{2}(p_1 - p_3)$$

lies on the boundary of $K_1$; see (3.9). In the same way one can prove that $q_2' = p_2 + \frac{1}{2}(p_3 - p_1) \in$ bd $K_2$, $q_3' = p_4 + \frac{1}{2}(p_3 - p_1) \in$ bd $K_3$, and $q_4' = p_4 + \frac{1}{2}(p_1 - p_3) \in$ bd $K_4$. Moreover, the points $q_1', \ldots, q_4'$ form a parallelogram. We may assume that the intersection point of the diagonals of

## 3.3. REGULAR 4-COVERINGS

$Q$ coincides with the origin 0. Thus we get

$$p_2 - p_3 = 2\lambda q_1, \quad p_2 - p_1 = 2\lambda q_2. \tag{3.10}$$

Let $\overline{K} = \frac{1}{2}(p_1 + p_3) + 2\lambda K$. Since $q_1 \in \operatorname{bd} K$ we have that

$$\frac{1}{2}(p_1 + p_3) + 2\lambda q_1 = \frac{1}{2}(p_1 + p_3) + (p_2 - p_3) = p_2 + \frac{1}{2}(p_1 - p_3) = q'_1$$

belongs to the boundary of $\overline{K}$. Since the points $q_2, q_3,$ and $q_4$ also belong to bd $K$, we obtain that $q'_2, q'_3, q'_4 \in \operatorname{bd} \overline{K}$.

Now we shall prove that $\cup_{i=1}^{4} K_i \subset \overline{K}$. From (3.7), (3.8), and (3.10) we have

$$K_1 = (1-\lambda)t_1 + \lambda K = p_2 - \lambda q_2 + \lambda K = \tag{3.11}$$

$$p_2 - \tfrac{1}{2}(p_2 - p_1) + \lambda K = \tfrac{1}{2}(p_1 + p_2) + \lambda K.$$

Let $u \in K_1$, say. Then there exists a point $v \in K$ such that $u = \frac{1}{2}(p_1+p_2)+\lambda v$, by (3.11). The convexity of $K$ implies that $v' = \frac{1}{2}(q_1+v) \in K$. Therefore the point $u' = \frac{1}{2}(p_1+p_3)+2\lambda v'$ belongs to $\overline{K}$. Since

$$u - u' = \frac{1}{2}(p_1 + p_2) + \lambda v - (\frac{1}{2}(p_1 + p_3) + \lambda(q_1 + v)) = \frac{1}{2}(p_2 - p_3) - \lambda q_1 = 0,$$

we obtain that $u \in \overline{K}$.

Assume that there exist a positive number $\lambda' < \lambda$ and a point $y$ such that $K^* = y + 2\lambda' K$ contains $\cup_{i=1}^{4} K_i$. Our aim is to obtain a contradiction. Since $q'_i \in K_i$, we have $q'_1, \ldots, q'_4 \in K^*$. This means that there exist four points $q''_1, \ldots, q''_4 \in K$ with

$$q'_i = y + 2\lambda' q''_i, \quad i = 1, \ldots, 4. \tag{3.12}$$

Clearly, these four points are vertices of a parallelogram. Moreover, the equations (3.12) imply

$$2\lambda'(q''_1 - q''_2) = p_1 - p_3 = \tfrac{1}{\mu}(q_1 - q_2) \quad \text{and} \tag{3.13}$$

$$2\lambda'(q''_2 - q''_3) = p_2 - p_4 = \tfrac{1}{\mu}(q_2 - q_3).$$

Since $2\lambda'\mu < 2\lambda\mu = 1$, we have

$$\|q''_1 - q''_2\|_E > \|q_1 - q_2\|_E \quad \text{and} \quad \|q''_2 - q''_3\|_E > \|q_2 - q_3\|_E, \tag{3.14}$$

where $\|\cdot\|_E$ is the usual Euclidean norm. Let $[y_1, y_2]$ be an affine diameter of $K$ parallel to $L(q_1, q_2)$. Then, obviously, the Euclidean lengths of all chords of $K$ parallel to $[y_1, y_2]$ and lying in the same half-plane with respect to the affine hull of $[y_1, y_2]$ monotonously decrease when their distance to $[y_1, y_2]$ increases. Therefore it is impossible that there exists a parallelogram with vertices $q''_1, \ldots, q''_4$ from $K$ with the properties (3.13) and (3.14). □

The above theorem refers to *all* convex bodies not necessarily being centrally symmetric. But if we restrict ourselves to centrally symmetric convex bodies, we can better clarify how the four homothets are placed to each other, as well as with respect to the original body. First we prove two lemmas. The first one can easily obtained from the monotonicity lemma, and we omit its proof.

**Lemma 3.3.3.** *Let $\mathcal{C}$ be the unit circle in a normed plane $(\mathbb{M}^2, \|\cdot\|)$, and $[p_1, q_1]$, $[p_2, q_2]$ be parallel chords of $\mathcal{C}$ having the same length in the norm such that the origin o does not belong to the open half-plane determined by $L(p_1, q_1)$ and containing $p_2$ and $q_2$. If $p_1 \prec p_2 \prec q_2 \prec q_1$, then the segments $[p_2, \frac{1}{\|p_1-q_2\|}(p_1-q_2)]$, $[q_2, \frac{1}{\|q_1-p_2\|}(q_1-p_2)]$ belong to $\mathcal{C}$ and contain the points $p_1$ and $q_1$, respectively.*

**Lemma 3.3.4.** *Let $\mathcal{C}$ be the unit circle of a normed plane $(\mathbb{M}^2, \|\cdot\|)$, and $P$ be a parallelogram inscribed in $\mathcal{C}$ whose diagonals do not intersect at the origin. Then two opposite sides of $P$ are segments on the unit circle $\mathcal{C}$.*

*Proof.* Let $p_1, p_2, p_3, p_4$ be the vertices of $P$, i.e., we have $\|p_1 - p_2\| = \|p_4 - p_3\|$. Consider the case when the origin 0 does not belong to the open half-plane determined by $L(p_1, p_2)$ and containing $p_3$ and $p_4$. Lemma 3.3.3 implies that the segments $[p_2, p_3]$ and $[p_4, p_1]$ are contained in $\mathcal{C}$. Let now the origin 0 belong to the open half-plane determined by $L(p_1, p_2)$ and containing $p_3$ and $p_4$. Then, for the parallelogram with vertices $-p_1, p_3, p_4, -p_2$, we apply the above case and obtain that the segments $[-p_1, p_3]$ and $[-p_2, p_4]$ belong to $\mathcal{C}$. Since $L(p_3, p_4)$ is parallel to $L(-p_1, -p_2)$, only the following relations are possible: $p_1 \prec p_2 \prec -p_1 \prec p_3 \prec p_4 \prec -p_2$ or $p_1 \prec p_2 \prec p_3 \prec -p_1 \prec -p_2 \prec p_4$. On the other hand, $L(p_1, -p_2) \parallel L(p_2, -p_1)$ and $L(p_1, p_4) \parallel L(p_2, p_3)$. Therefore the convexity of $\mathcal{C}$ implies that the points $p_1, p_4, -p_2$ as well as the points $p_2, p_3, -p_1$ are collinear. Thus, again the convexity of $\mathcal{C}$ yields that $[p_1, p_4]$ and $[p_2, p_3]$ are contained in $\mathcal{C}$. □

**Lemma 3.3.5.** *Let $\mathrm{cov}(P, Q, \lambda)$ be a regular 4-covering of the unit disc $\mathcal{D}$ of the normed plane $(\mathbb{M}^2, \|\cdot\|)$. Then the diagonals of $P$ and $Q$ intersect at the origin 0.*

*Proof.* Let $p_1, p_2, p_3, p_4$ be the vertices of $P$, $q_1, q_2, q_3, q_4$ be the vertices of $Q$, and $p_1 \prec q_1 \prec \ldots \prec p_4 \prec q_4$. Assume that neither the diagonals of $P$ nor the diagonals of $Q$ intersect at the origin. Then, by Lemma 3.3.4, two opposite sides of $P$ are segments on the unit circle $\mathcal{C}$, say $[p_1, p_2]$ and $[p_3, p_4]$. Again by Lemma 3.3.4, $[q_1, q_2]$ or $[q_1, q_4]$ is a segment on $\mathcal{C}$. If $[q_1, q_2]$ is that segment, then the points $p_1, q_1, p_2, q_2$ are collinear. But $P$ and $Q$ are quasi-dual, i.e., $L(q_1, q_2) \parallel L(p_1, p_3)$. This contradicts the fact that $P$ is a parallelogram. Therefore the diagonals of at least one of the parallelograms $P$ or $Q$ intersect at the origin. Suppose that the diagonals of $Q$ do not intersect at the origin. Then Lemma 3.3.4 implies again that $[q_1, q_2]$

and $[q_3, q_4]$ (or $[q_2, q_3]$ and $[q_4, q_1]$) belong to $\mathcal{C}$. Thus we get $p_2 \in [q_1, q_2]$ and $p_4 \in [q_3, q_4]$ (or $p_1 \in [q_4, q_1]$ and $p_3 \in [q_2, q_3]$). Since $L(p_2, p_4) \parallel L(q_2, q_3)$ (or $L(p_1, p_3) \parallel L(q_1, q_2)$), by Lemma 3.3.2 and Lemma 3.3.1 we obtain $\mu = 1$, which is impossible. Analogously, if the diagonals of $P$ do not intersect at the origin, then $\lambda = 1$, which is also impossible. □

**Theorem 3.3.2.** *In a normed plane, let $\cup_{i=1}^{4} \mathcal{D}(x_i, \lambda)$ be a regular 4-covering of a disc $\mathcal{D}(x, \mu)$ derived from the quasi-dual parallelograms with vertices $p_1, p_2, p_3, p_4$ and $q_1, q_2, q_3, q_4$; see again Figure 3.5. Then, for $i = 1, \ldots, 4$ (and $x_5 \equiv x_1$, $p_5 \equiv p_1$), we have*

*(i) $\{x, p_i\} \in \mathcal{C}(x_i, \lambda) \cap \mathcal{C}(x_{i+1}, \lambda)$;*

*(ii) $\mathcal{C}(x_1, \lambda)$ and $\mathcal{C}(x_3, \lambda)$ touch each other, as also $\mathcal{C}(x_2, \lambda)$ and $\mathcal{C}(x_4, \lambda)$ do;*

*(iii) $x_i$ is the midpoint of $[p_i, p_{i+1}]$;*

*(iv) the points $x_1, x_2, x_3, x_4$ form a parallelogram $P'$ of side-length $\mu$ which is inscribed in $\mathcal{C}(x, \mu)$, and a parallelogram $Q'$ exists which is also inscribed in $\mathcal{C}(x, \mu)$ such that $P'$ and $Q$ are quasi-dual.*

*Proof.* Without loss of generality we assume that $x \equiv 0$. If $t_i, i = 1, \ldots, 4$, are determined as in (3.5), then the homothety $\varphi_i$ with center $t_i$ and ratio $\frac{\lambda}{\mu}$ maps $\mathcal{D}(0, \mu)$ into $\mathcal{D}(x_i, \lambda)$. Since $x_i = \varphi_i(0)$, we have $x_i = (1 - \lambda)t_i$. On the other hand, the origin 0 is the midpoint of $[q_2, q_4]$ and $\varphi_1(\{q_4, q_2\}) = \{p_1, p_2\}$. Thus we see that $x_1$ is the midpoint of $[p_1, p_2]$. By Lemma 3.3.1 we have $\|p_1 - p_2\| = 2\lambda$. From here we get $\|x_1\| = \lambda$. In the same way one can prove that $\|x_2\| = \|x_3\| = \|x_4\| = \lambda$. The last statement of Theorem 3.3.2 is evident. □

### 3.3.2 A lattice covering of the plane based on a regular 4-covering

A *lattice* of vectors in $\mathbb{M}^2$ is the collection $\mathbb{L} = \mathbb{L}(u, v)$ of integer-coefficient linear combinations of a pair of linearly independent vectors $u$ and $v$. The pair $\{u, v\}$ is called the *basis* of the lattice $\mathbb{L}$, and the parallelogram spanned by $u$ and $v$ is said to be the respective *basis parallelogram*. A *lattice covering* of the plane is a covering of the plane whose members are translates of a given convex body, where the translation vectors are taken from the lattice. A covering of the plane with circles of radius $\lambda$ has *margin* $\mu$, $0 \leq \mu \leq \lambda$, provided the plane remains covered if any circle of the covering is replaced by the concentric circle of radius $\lambda - \mu$.

Let $\cup_{i=1}^{4} \mathcal{D}(x_i, \lambda)$ be a regular 4-covering of the unit disc $\mathcal{D}$ of a strictly convex normed plane. If $\mathbb{L} = \mathbb{L}(x_2 - x_1, x_4 - x_1)$, then $\mathcal{D}(x_1, \lambda) + \mathbb{L}$ is a lattice covering of the plane. We call this covering a *regular 4-covering of the plane generated by* $\cup_{i=1}^{4} \mathcal{D}(x_i, \lambda)$. From Theorem 3.3.2 we immediately obtain the following properties.

**Proposition 3.3.1.** *Let $\mathcal{D}(x, \lambda) + \mathbb{L}$ be a regular 4-covering of the plane, where $\mathcal{D}(x, \lambda)$ is a circle in a strictly convex normed plane and $\mathbb{L} = \mathbb{L}(u, v)$. Then*

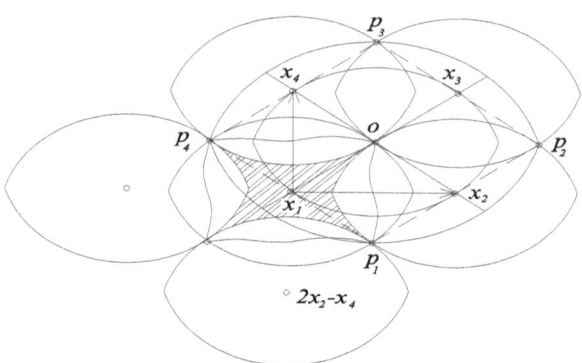

Figure 3.5: A regular 4-covering of a strictly convex normed plane.

(i) the margin of $\mathcal{D}(x,\lambda) + \mathbb{L}$ is zero;

(ii) $\mathcal{D}(x,\lambda) + \mathbb{L}(u+v, u-v)$ and $\mathcal{D}(u+x,\lambda) + \mathbb{L}(u+v, u-v)$ are packings of the plane.

Let $\mathcal{D}(x,\lambda) + \mathbb{L}$ be a lattice covering of the plane, where $\mathcal{D}(x,\lambda)$ is a disc of a strictly convex normed plane. The *Voronoi region of* $\mathcal{D}(x,\lambda)$ is the set of points whose distances from $x$ do not exceed the distance from the center of any other disc of the covering $\mathcal{D}(x,\lambda) + \mathbb{L}$. The gray area of $\mathcal{D}(x,\lambda)$ is the closure of the set of points in $\mathcal{D}(x,\lambda)$ which belong to no other disc of the covering $\mathcal{D}(x,\lambda)$. Various properties of Voronoi regions and of gray areas of discs $\mathcal{D}(x,\lambda)$ in the Euclidean subcase are given in [13]. Here we give some properties of the Voronoi region and the gray area of $\mathcal{D}(x,\lambda)$ if $\mathcal{D}(x,\lambda) + \mathbb{L}$ is a regular 4-covering of a strictly convex normed plane.

**Proposition 3.3.2.** *Let* $\mathcal{D}(x_1,\lambda) + \mathbb{L}$ *be a regular 4-covering of a strictly convex normed plane generated by* $\cup_{i=1}^{4} \mathcal{D}(x_i, \lambda)$; *see Figure 3.5. If* $\mathcal{V}$ *is the Voronoi region of* $\mathcal{D}(x_1,\lambda) + \mathbb{L}$ *and* $\mathcal{G}$ *is its gray area, then*

(i) $\mathcal{G} \subset \mathcal{V} \subset \mathcal{D}(x_1,\lambda)$;

(ii) $\mathcal{V}$ and $\mathcal{G}$ are symmetric with respect to $x_1$;

(iii) the family of translates of $\mathcal{V}$, obtained by the basis vectors of the lattice $\mathbb{L}$, is a tiling of the plane;

(iv) the Voronoi region $\mathcal{V}$ and the gray area $\mathcal{G}$ are inscribed in $\mathcal{C}(x_1,\lambda)$;

(v) the boundary of the gray area $\mathcal{G}$ is the union of four circular arcs of the covering;

*(vi) the convexity of $\mathcal{V}$ implies that it is parallelogram.*

*Proof.* Let $\mathcal{C}(x_i, \lambda) \cap \mathcal{C}(x_{i+1}, \lambda) = \{0, p_i\}$. The inclusions $\mathcal{G} \subset \mathcal{V}$ and $\mathcal{V} \subset \mathcal{D}(x_1, \lambda)$ follow from the fact that the bisector of $[x_1, x_2]$ between the points $0$ and $p_1$, say, belongs to $\mathcal{D}(x_1, \lambda) \cap \mathcal{D}(x_2, \lambda)$, and also that $0, p_1 \in B(x_1, x_2)$ implies (iv). Since, in view of Theorem 3.3.2, we have $\|x_1 - x_4\| = \frac{1}{2} \|p_1 - p_3\| = 1$, the distance between the centers of $\mathcal{C}(2x_1 - x_4, \lambda)$ and $\mathcal{C}(x_4, \lambda)$ equals 2, which is strictly larger than $2\lambda$. Therefore $\mathcal{C}(2x_1 - x_4, \lambda)$ and $\mathcal{C}(x_4, \lambda)$ have no points in common. Thus also (v) is proved. In order to prove (vi), we note that every line parallel to $[x_1, x_2]$ intersects $B(x_1, x_2)$ in exactly one point; see [72, Proposition 15]. If this point belongs to the part of $B(x_1, x_2)$ between $0$ and $p_1$, then it lies neither in the open half-plane bounded by $L(0, p_1)$ and containing $x_1$, nor in its opposite half-plane. Therefore, this part is a segment, and the proof of (vi) is done. The statements (ii) and (iii) are evident. □

## 3.4 Configurations of Minkowskian circles related to covering problems

### 3.4.1 Configurations of Minkowskian circles related to a regular 4-covering

Let $(\mathbb{M}^2, \|\cdot\|)$ be a strictly convex normed plane, and $\mathcal{C}(x_i, \lambda)$, $i = 1, \ldots, 4$, be four circles passing through a point $p$ such that $\mathcal{C}(x_i, \lambda)$ and $\mathcal{C}(x_{i+1}, \lambda)$ do not touch each other, where $x_5 \equiv x_1$. Then $\mathcal{C}(x_i, \lambda)$ and $\mathcal{C}(x_{i+1}, \lambda)$ have exactly one second intersection point, denoted by $p_{i+1}$. The next theorem clarifies what configuration is obtained if the second intersection points of these circles lie on one circle. More precisely, we have

**Theorem 3.4.1.** *In a strictly convex normed plane, let there be given four circles $\mathcal{C}(x_i, \lambda)$, $i = 1, \ldots, 4$, passing though a point $p$ such that $\mathcal{C}(x_i, \lambda)$ and $\mathcal{C}(x_{i+1}, \lambda)$ do not touch each other, whereas $\mathcal{C}(x_i, \lambda)$ and $\mathcal{C}(x_{i+2}, \lambda)$ touch each other ($x_5 \equiv x_1$, $x_6 \equiv x_2$). If $p_{i+1}$ ($p_5 \equiv p_1$) is the second intersection point of $\mathcal{C}(x_i, \lambda)$ and $\mathcal{C}(x_{i+1}, \lambda)$, and $p_1, p_2, p_3, p_4$ lie on the same circle of radius $\mu > \lambda$, then $\cup_{i=1}^{4} \mathcal{D}(x_i, \lambda)$ is a regular 4-covering of $\mathcal{D}(p, \mu)$.*

*Proof.* Without loss of generality we can assume that $p \equiv 0$. Consider the circles $\mathcal{C}(0, \mu)$ and $\mathcal{C}(x_i, \lambda)$. Then we have $\mathcal{C}(0, \mu) \cap \mathcal{C}(x_i, \lambda) = \{p_i, p_{i+1}\}$. Thus Lemma 1.2.5 implies that if $\gamma$ is the arc of $\mathcal{C}(0, \mu)$ between $p_i$ and $p_{i+1}$ that does not contain the remaining points of the set $\{p_1, \ldots, p_4\}$, and $\gamma'$ is the arc of $\mathcal{C}(x_i, \lambda)$ with endpoints $p_i, p_{i+1}$ that does not contain $0$, then $\gamma \in \operatorname{conv} \gamma'$. This means that $\cup_{i=1}^{4} \mathcal{D}(x_i, \lambda)$ is a covering of $\mathcal{D}(0, \mu)$. The property that the circles $\mathcal{C}(x_i, \lambda)$ and $\mathcal{C}(x_{i+2}, \lambda)$ touch each other implies that $x_1, x_2, x_3, x_4$ form a parallelogram whose diagonals intersect at $0$ (cf. Lemma 1.2.6). Thus, by Lemma 1.2.7, we get

$$x_1 + x_2 = p_2, \quad x_2 + x_3 = p_3,$$
$$x_3 + x_4 = p_4, \quad x_4 + x_1 = p_1;$$

whence

$$\begin{aligned} p_1 - p_2 &= x_4 - x_2, \quad p_2 - p_3 = x_1 - x_3, \\ p_2 + p_4 &= x_1 + x_2 + x_3 + x_4, \quad p_1 + p_3 = x_1 + x_2 + x_3 + x_4. \end{aligned} \quad (3.15)$$

Therefore the points $p_1, p_2, p_3, p_4$ form a parallelogram with diagonals intersecting at 0. Moreover, its sides are parallel to the diagonals of the parallelogram obtained by $x_1, x_2, x_3, x_4$. The equations (3.15) imply that $\|p_2 - p_3\| = \|p_1 - p_2\| = 2\lambda$. Thus we obtain that $x_{i+1}$ is the midpoint of $[p_i, p_{i+1}]$. If $\mathrm{R}^+_{x_i}(0) \cap \mathcal{C}(0, \lambda) = \{q_i\}$, then the points $q_1, q_2, q_3, q_4$ form a parallelogram which is quasi-dual to that with vertices $p_1, p_2, p_3, p_4$. Now it is easy to see that $\cup_{i=1}^{4} \mathcal{D}(x_i, \lambda)$ is a regular 4-covering of $\mathcal{D}(0, \mu)$. □

### 3.4.2 Miquel configurations of circles of equal radii

More general than in the previous subsection, we consider configurations $\{\mathcal{C}(x_i, \lambda), i = 1, \ldots, 4\}$ in which the first intersection points of $\mathcal{C}(x_i, \lambda)$ and $\mathcal{C}(x_{i+1}, \lambda)$, with $x_5 \equiv x_4$, do not coincide. For such configurations in strictly convex normed planes the next theorem holds. This theorem was also proved by Asplund and Grünbaum in [6], but under the additional assumption that the plane be smooth. In the Euclidean subcase this theorem is known as *Miquel's Six-Circles Theorem*.

**Theorem 3.4.2.** *In a strictly convex normed plane* $(\mathbb{M}^2, \|\cdot\|)$, *let there be given four circles* $\mathcal{C}(x_i, \lambda)$, $i = 1, \ldots, 4$, *such that* $\mathcal{C}(x_i, \lambda)$ *and* $\mathcal{C}(x_{i+1}, \lambda)$, *with* $x_5 \equiv x_1$, *have exactly two intersection points* $p_{i+1}$ *and* $q_{i+1}$, *where* $p_5 \equiv p_1$ *and* $q_5 \equiv q_1$. *Then the points* $p_1, \ldots, p_4$ *lie on the same circle of radius* $\lambda$ *if and only if* $q_1, \ldots, q_4$ *lie on the same circle of radius* $\lambda$.

*Proof.* Without loss of generality we assume that $p_i \in \mathcal{C}(o, \lambda)$, $i = 1, \ldots, 4$. By Lemma 1.2.7 we have

$$\begin{aligned} x_1 + x_2 &= p_2 + q_2 & x_1 &= p_1 + p_2 \\ x_2 + x_3 &= p_3 + q_3 & \text{and} \quad x_2 &= p_2 + p_3 \\ x_3 + x_4 &= p_4 + q_4 & x_3 &= p_3 + p_4 \\ x_4 + x_1 &= p_1 + q_1 & x_4 &= p_4 + p_1. \end{aligned} \quad (3.16)$$

The equations (3.16) imply that $q_i = p_i + p_{i+1} + p_{i+3} = p_1 + p_2 + p_3 + p_4 - p_{i+2}$, where $i = 1, \ldots, 4$ and $p_6 = p_2, p_7 = p_3$. Since $\|p_i\| = \lambda$, we get $\|p_1 + p_2 + p_3 + p_4 - q_i\| = \lambda$, which means that $q_i \in C(p_1 + p_2 + p_3 + p_4, \lambda)$. □

## 3.4. CONFIGURATIONS OF CIRCLES RELATED TO COVERING PROBLEMS

Now we consider a configuration of circles in an arbitrary normed plane a configuration of circles which is described in the above theorem. Let $p_1, \ldots, p_8$ be eight points. To every point $p_i$, $i = 1, \ldots, 8$, we assign a vertex of a cube. Consider the six quadruples of points that correspond to the vertices of each facet of the cube, e.g.,

$$(p_1, p_2, p_3, p_4), (p_1, p_2, p_5, p_6), (p_2, p_3, p_7, p_6),$$
$$(p_3, p_4, p_8, p_7), (p_1, p_4, p_8, p_5), (p_5, p_6, p_7, p_8). \tag{3.17}$$

If five of the quadruples in (3.17) are concyclic, then this configuration is called a *Miquel configuration*. If for four points there exists a circle containing them, we say that these four points form a *concyclic quadruple*. Thus, Theorem 3.4.2 can be rewritten as follows: in a strictly convex normed plane, in any Miquel configuration of circles with equal radii *all six* quadruples are concyclic. Such a configuration of circles is called $(8_3, 6_4)$-*configuration* because it is formed by 8 points and 6 circles, any of these points lying on 3 circles, and any circle passing through 4 of these points. Theorem 3.4.2 can be extended to *all* normed spaces, but we need some preliminaries. According to Theorem 2.4 in [8] (see also [72, Proposition 22]) the intersection $\mathcal{I}$ of two circles $\mathcal{C}(p, \lambda)$ and $\mathcal{C}(q, \mu)$ in $(\mathbb{M}^2, \|\cdot\|)$ can only have the following forms:

(i) $\mathcal{I} = \emptyset$;

(ii) $\mathcal{I} = \mathcal{C}(p, \lambda) = \mathcal{C}(q, \mu)$;

(iii) $\mathcal{I}$ consists of two closed, disjoint segments (one of them or both may be reduced to a point) lying on the opposite sides of the line $G$ through $p$ and $q$;

(iv) $\mathcal{I}$ consists of two segments (one of them or both may be reduced to a point) with common point $p_1$ or $p_2$, where $\{p_1, p_2\} = G \cap \mathcal{C}(p, \lambda)$.

Note that if $\lambda = \mu$ and $p \neq q$, (iii) or (iv) occur if and only if $\|p - q\| \leq 2\lambda$. If the plane is strictly convex, then the intersection of $\mathcal{C}(p, \lambda)$ and $\mathcal{C}(q, \lambda)$ consists of exactly two points if and only if $\|p - q\| < 2\lambda$. If $\|p - q\| = 2\lambda$, then $\mathcal{C}(p, \lambda) \cap \mathcal{C}(q, \lambda)$ consists of exactly one point.

In a normed plane, let the circles $\mathcal{C}(p, \lambda)$ and $\mathcal{C}(q, \lambda)$ intersect properly, and let their intersection consist of the segments $\mathcal{I}_1$ and $\mathcal{I}_2$ (possibly degenerate). If the point $r_1$ belongs to $\mathcal{I}_1$, we call the point $r_2 = p + q - r_1$ the *conjugate of* $r_1$ *with respect to the circles* $\mathcal{C}(p, \lambda)$ *and* $\mathcal{C}(q, \lambda)$. If the intersection of both circles consists only of one point (i.e., in (iv) the two segments are reduced to points that coincide), then this point is the conjugate of itself. Clearly, we have $r_2 \in \mathcal{I}_2$, and $r_1$ is the conjugate of $r_2$. If the plane is strictly convex and $\mathcal{C}(p, \lambda) \cap \mathcal{C}(q, \lambda) = \{r_1, r_2\}$, then $r_1$ and $r_2$ are conjugates of each other.

Let there be given a circle $\mathcal{C}(x, \lambda)$. Let the points $p_1, \ldots, p_4$ be placed on $\mathcal{C}(x, \lambda)$ in this order. Then the monotonicity lemma implies that $\|p_i - p_{i+1}\| \leq 2\lambda$ for $i = 1, \ldots, 4$ and

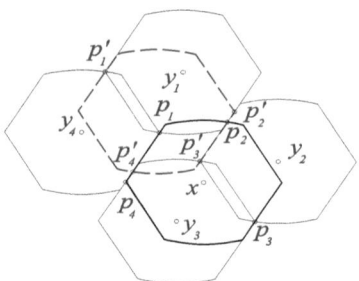

Figure 3.6: Miquel's Theorem

$p_5 = p_1$. Therefore the circles $\mathcal{C}(p_i, \lambda)$ and $\mathcal{C}(p_{i+1}, \lambda)$ intersect. Clearly, the point $x$ belongs to $\mathcal{C}(p_i, \lambda) \cap \mathcal{C}(p_{i+1}, \lambda)$ for every $i = 1, \ldots, 4$. Let $y_i$ be the conjugate point of $x$ with respect to $\mathcal{C}(p_i, \lambda)$ and $\mathcal{C}(p_{i+1}, \lambda)$. We call the configuration of circles

$$\mathcal{C}(y_1, \lambda \mathcal{C}), \ldots, \mathcal{C}(y_4, \lambda), \mathcal{C}(x + \lambda) \tag{3.18}$$

the *Miquel configuration induced by the points* $p_1, \ldots, p_4$.

**Remark 3.4.1.** *If (3.18) is the Miquel configuration induced by the points* $p_1, \ldots, p_4$ *which lie on* $\mathcal{C}(x, \lambda)$, *then we can speak about the conjugate point of* $p_{i+1}$ *with respect to the circles* $\mathcal{C}(y_i, \lambda)$ *and* $\mathcal{C}(y_{i+1}, \lambda)$, *where* $y_5 = y_1$. *Indeed, since* $y_i \in \mathcal{C}(p_i, \lambda) \cap \mathcal{C}(p_{i+1}, \lambda)$, *we have that* $p_i, p_{i+1} \in \mathcal{C}(y_i, \lambda)$, *which yields* $p_{i+1} \in \mathcal{C}(y_i, \lambda) \cap \mathcal{C}(y_{i+1}, \lambda)$.

**Theorem 3.4.3.** *In a normed plane* $(\mathbb{M}^2, \|\cdot\|)$ *with unit circle* $\mathcal{C}$, *let there be given a circle* $\mathcal{C}(x, \lambda)$. *Let the points* $p_1, \ldots, p_4$ *be placed in this order on* $\mathcal{C}(x, \lambda)$, *and let*

$$\mathcal{C}(y_1, \lambda), \ldots, \mathcal{C}(y_4, \lambda), \mathcal{C}(x, \lambda) \tag{3.19}$$

*be the Miquel configuration induced by the points* $p_1, \ldots, p_4$. *If* $p'_{i+1}$ *is the conjugate point of* $p_{i+1}$ *with respect to* $\mathcal{C}(y_i, \lambda)$ *and* $\mathcal{C}(y_{i+1}, \lambda)$, *then the points* $p'_i$ *lie on a circle of radius* $\lambda$, *where* $i = 1, \ldots, 4$, $p'_5 = p'_1$, $p_5 = p_1$, *and* $y_5 = y'_5$; *see Figure 3.6.*

*Proof.* Without loss of generality we assume that $x = 0$. Since the points $p_1, \ldots, p_4$ induce the Miquelian configuration (3.19), we have

$$y_i = p_i + p_{i+1}, \quad i = 1, \ldots, 4. \tag{3.20}$$

## 3.4. CONFIGURATIONS OF CIRCLES RELATED TO COVERING PROBLEMS

On the other hand, the points $p_{i+1}$ and $p'_{i+1}$ are conjugate with respect to $\mathcal{C}(y_i + \lambda)$ and $\mathcal{C}(y_{i+1}, \lambda)$. Therefore we get

$$p_{i+1} + p'_{i+1} = y_i + y_{i+1}, \quad i = 1, \ldots, 4. \tag{3.21}$$

Thus, from (3.20) and (3.21) we obtain

$$p'_{i+1} = y_i + y_{i+1} - p_{i+1} = p_i + p_{i+1} + p_{i+2} = p_1 + \ldots + p_4 - p_{i+3}, \tag{3.22}$$

where $p_6 = p_2$, $p_7 = p_3$. Hence $\|p'_{i+1} - (p_1 + \ldots + p_4)\| = \|p_{i+3}\| = \lambda$. □

**Remark 3.4.2.** *If the plane $(\mathbb{M}^2, \|\cdot\|)$ in Theorem 3.4.3 is strictly convex, then $p_{i+1}$ and $p'_{i+1}$ are the intersection points of $\mathcal{C}(y_i, \lambda)$ and $\mathcal{C}(y_{i+1}, \lambda)$. In such a case, Theorem 3.4.3 appears to be a reformulation of Theorem 3.4.2. Restricted to the strictly convex case, Theorem 3.4.3 is, however, more general than Theorem 3.4.2. If the points $p_1$ and $p_3$ are opposite with respect to $\mathcal{C}(x, \lambda)$, then from (3.20) we have $\|y_1 - y_2\| = 2\lambda$ and $\|y_3 - y_4\| = 2\lambda$ (under the assumption that $x = 0$). This means that $\mathcal{C}(y_1, \lambda)$ and $\mathcal{C}(y_2, \lambda)$, as well as $\mathcal{C}(y_3, \lambda)$ and $\mathcal{C}(y_4, \lambda)$, have exactly one common point. In other words, if two pairs of circles corresponding to the quadruples in Theorem 3.4.2 have only one point in common, Theorem 3.4.2 is also true.*

### 3.4.3 Miquel configurations of circles having arbitrary radii

In this subsection we consider Miquel configurations consisting of circles of arbitrary radii. It is our aim to ascertain whether the theorem of Miquel holds for such configurations. A crucial role in our investigations plays the fact that any strictly convex, smooth normed plane is a Möbius plane; this is proved Section 3.1, see Theorem 3.1.1 there. For our purpose we need more facts about Möbius planes.

Let $\Sigma = (\mathfrak{P}, \mathfrak{C})$ be a Möbius plane. A set of generalized circles having a unique common generalized point form a *parabolic bundle*. If the following statement (F) holds in a Möbius plane $\Sigma$, then $\Sigma$ is called an *(F)-plane*[1]:

(F) Every generalized circle that touches three different generalized circles of a parabolic bundle belongs to the same bundle.

It is easy to check that every strictly convex, smooth normed plane is an (F)-plane. For a Möbius plane the notions of *concyclic quadruple of generalized points* and *a Miquelian configuration of generalized circles* are defined as in the previous subsection, i.e., the generalized points

---
[1] Coming from the German word "Fährte".

$p_1, p_2, p_3, p_4 \in \mathfrak{P}$ are said to be *concyclic* if there exists a generalized circle $C \in \mathfrak{C}$ containing them. The generalized points $p_1, \ldots p_8 \in \mathfrak{P}$ form a Miquelian configuration if five of the quadruples

$$(p_1, p_2, p_3, p_4), (p_1, p_2, p_5, p_6), (p_2, p_3, p_7, p_6),$$
$$(p_3, p_4, p_8, p_7), (p_1, p_4, p_8, p_5), (p_5, p_6, p_7, p_8). \tag{3.23}$$

are concyclic. If in any Miquelian configuration in a Möbius plane $(\mathfrak{P}, \mathfrak{C})$ all six quadruples are concyclic, then $(\mathfrak{P}, \mathfrak{C})$ is said to be a *Miquelian Möbius plane*.

Let $\Sigma$ and $\Sigma'$ be two Möbius planes. If there exists a one-to-one correspondence $\sigma : \Sigma \to \Sigma'$ mapping concyclic generalized points into concyclic generalized points, and non-concyclic generalized points into non-concyclic ones, then $\Sigma$ and $\Sigma'$ are called *isomorphic* and $\sigma$ is said to be a *homography from $\Sigma$ to $\Sigma'$*. Clearly, the isomorphism between Möbius planes is an equivalence relation. Note also that if a Möbius plane is an (F)-plane, then all planes isomorphic to it are (F)-planes, too.

We mention that a class of Möbius planes can be constructed in an algebraic way. Let $\mathbb{F}$ and $\mathbb{E} \supset \mathbb{F}$ be commutative fields and $[\mathbb{E} : \mathbb{F}] = 2$. One can consider the elements of $\mathbb{E} \cup \{\infty\}$, where $\infty$ is a formal symbol, as generalized points and define generalized circles (usually called *chains*) as sets

$$\{x \in \mathbb{E} \cup \{\infty\} | \frac{p-r}{q-r} : \frac{p-x}{q-x} \in \mathbb{F} \cup \infty\},$$

where $p, q$, and $r$ are three pairwise different points from $\mathbb{E}$. Then the so-defined incidence structure is a Möbius plane (see [11, § 2]) and we denote it by $\text{Mo}(\mathbb{F}, \mathbb{E})$. Note also that in the classical case $\mathbb{F} = \mathbb{R}$, $\mathbb{E} = \mathbb{C}$, and $\text{Mo}(\mathbb{F}, \mathbb{E})$ is the inversive plane.

Let now $(\mathbb{M}^2, \|\cdot\|)$ be a strictly convex, smooth normed plane, and consider this plane as a Möbius plane $\Sigma = (\mathfrak{P}, \mathfrak{C})$. A homography $\varphi$ in $\Sigma$ that is involutory and leaves the generalized points of a generalized circle $C$ fixed such that no other generalized point is fixed is called *inversion with respect to the generalized circle $C$*. It is clear that such a homography exists at least for the Euclidean subcase.

**Proposition 3.4.1.** *In a strictly convex, smooth normed plane, let $\varphi$ be an inversion with respect to the circle $C$ with center $x$. Then $\varphi(x) = \infty$.*

*Proof.* Assume that $\varphi(x) = x' \neq \infty$. Clearly, $x \neq x'$. Let $\text{L}(x, x') \cap C = \{p_1, p_2\}$ and $\varphi(\infty) = y$. Then $y \neq x$ and $y \neq \infty$. We distinguish the following cases:

(1) $y \notin \text{L}(x, x')$. Then the image of $\text{L}(p_1, p_2)$ is a circle passing through $p_1, p_2$, and $x'$, a contradiction.

## 3.4. CONFIGURATIONS OF CIRCLES RELATED TO COVERING PROBLEMS

(2a) $y \in L(x, x')$ and $y$ is interior with respect to $C$. Consider the circle $C_1$ with center $\frac{x+y}{2}$ and radius $\frac{\|x-y\|}{2}$. Clearly, $C_1 \cap C = \emptyset$. On the other hand, $\varphi(C_1) = C_1'$ is a line through $x'$. If the point $x'$ is interior with respect to $C$, then $C_1' \cap C \neq \emptyset$. This is not true, and thus we obtain that $x'$ is exterior with respect to $C$. Let $G$ be a supporting line of $C$ through $x'$, and let $G \cap C = \{z\}$. If $\varphi(G) = C_2$, then $C_2$ is the circle through $x, y, z$, and $C_2 \cap C = \{z\}$. Denote by $x_2$ the center of $C_2$. Since $(\mathbb{M}^2, \|\cdot\|)$ is strictly convex, the point $x_2$ lies on $L(x, z)$. Therefore $x_2$ is the midpoint of the segment $[x, z]$. If the line $H$ supports $C_2$ at $x$, then $H$ is parallel to $G$, by the fact that $(\mathbb{M}^2, \|\cdot\|)$ is smooth. Let $H \cap C = \{q_1, q_2\}$. If $\varphi(H) = C_3$, then $C_3$ is the circle through $q_1, q_2, x'$. Moreover, $G$ supports $C_3$ at $x'$ and $C_2 \cap C_3 = \emptyset$. If $q_2$ lies in the half plane $\text{HS}_{x'}^+(x, z)$, the circular arc of $C_3$ between $q_1$ and $x'$ has to intersect $C_2$, a contradiction.

(2b) $y \in L(x, x')$ and $y$ is exterior with respect to $C$. Let $C_1$ be the circle with center $\frac{x+y}{2}$ and radius $\frac{\|x-y\|}{2}$, and let $C_1 \cap C = \{q_1, q_2\}$. According to [8, Theorem 2.4] (cited also in Subsection 3.4.2), the points $q_1$ and $q_2$ lie in the different half-planes with respect to $L(x, y)$. Note that the intersection point of $L(q_1, q_2)$ and $L(x, y)$ is $x'$, since $\varphi(C_1) = L(q_1, q_2)$. Thus we get that $x'$ is interior with respect to $C$. Consider the circle $C_2$ through $x'$ and $y$ with center $\frac{x'+y}{2}$. If $G$ supports $C_2$ at $x'$, the points $x$ and $y$ lie on opposite sides of $G$. Let $C_2 \cap C = \{t_1, t_2\}$. Then the segment $[t_1, t_2]$ has to intersect $L(x, y)$. Denote the respective intersection point by $t$. Since $\varphi(C_2) = L(t_1, t_2)$, the point $t$ has to coincide with $x$. Due to the fact that $t$ and $x$ lie in different half-planes with respect to $G$, again we get a contradiction. □

**Remark 3.4.3.** In [96], Stiles defined the inversion with respect to the unit circle $C$ of a normed plane as a mapping $\varphi$ of $\mathbb{M}^2 \setminus \{0\}$ onto itself that maps a point $x \neq 0$ onto the point $\frac{1}{\|x\|^2}x$. He proved that if the inversive image of some line is a circle, then $(\mathbb{M}^2, \|\cdot\|)$ is Euclidean. Theorem 3.4.5 below shows that Stiles' definition of inversion and ours are only equivalent in the Euclidean case.

For the proof of Theorem 3.4.5 we also need the following characterization of the Euclidean plane, proved in [9].

**Theorem 3.4.4.** *A normed plane with unit circle $C$ is Euclidean if and only if*
$$x, y \in C, \ \inf\{\|\alpha x + (1-\alpha)y\| : 0 \leq \alpha \leq 1\} = \frac{1}{2} \implies x + y \in C.$$

**Theorem 3.4.5.** *Let $(\mathbb{M}^2, \|\cdot\|)$ be a strictly convex, smooth normed plane, and let there exist the inversion $\varphi$ with respect to some circle of $(\mathbb{M}^2, \|\cdot\|)$. Then the plane is Euclidean.*

*Proof.* Without loss of generality we can assume that in $(\mathbb{M}^2, \|\cdot\|)$ there exists an inversion with respect to the unit circle $C$ of $(\mathbb{M}^2, \|\cdot\|)$. Let $p_1, p_2$ be points on $C$ and $x$ be a point on $[p_1, p_2]$ such that the segment $[p_1, p_2]$ supports $\frac{1}{2}C$ at $x$. Let $H$ be the supporting line of $C$

at $2x$. The smoothness of $(\mathbb{M}^2, \|\cdot\|)$ implies that $H$ is parallel to $L(p_1, p_2)$. Since $\varphi(\infty) = 0$, the inverse image $H'$ of $H$ is a circle through $0$. On the other hand, $\varphi(2x) = 2x$. Therefore the circle $H'$ passes through $2x$. Further on, the plane $(\mathbb{M}^2, \|\cdot\|)$ is strictly convex, and this means that $2x$ is the unique common point of $\mathcal{C}$ and $H$. Hence there do not exist common points of $H$ and $H'$ except for $2x$, and the line $H$ appears to be supporting $H'$ at $2x$. Thus the strict convexity of $(\mathbb{M}^2, \|\cdot\|)$ implies that the center of $H'$ lies on $L(0, 2x)$. But $0, 2x \in H'$, and therefore $H' = x + \frac{1}{2}\mathcal{C}$. Let $L(p_1, p_2) \cap (x + \frac{1}{2}\mathcal{C}) = \{q_1, q_2\}$, $R_{q_1}^+(0) \cap H = \{q_1'\}$, and $R_{q_2}^+(0) \cap H = \{q_2'\}$. For the points $q_1'$ and $q_2'$ the equations

$$\|2x - q_1'\| = 1, \quad \|2x - q_2'\| = 1 \tag{3.24}$$

hold. Since the inversive image of any line through $0$ is the same line, we have $\varphi(q_1, q_2) = \{q_1', q_2'\}$. Consider $G' = \varphi(G)$, where $G = L(p_1, p_2)$. Clearly, $G'$ is a circle through $0, p_1, p_2, q_1', q_2'$. Thus we obtain that $G' = 2x + \mathcal{C}$; see (3.24). Therefore the quadrangle with vertices $0, p_1, 2x$, and $p_2$ is a metric parallelogram (i.e., a quadrangle with opposite sides of equal lengths). But in a strictly convex normed plane every metric parallelogram is a parallelogram (see [72, Proposition 12]), i.e., we get that $x$ is the midpoint of $[p_1, p_2]$. Thus, in view of Theorem 3.4.4 the proof is complete. $\square$

**Theorem 3.4.6.** *If Miquel's theorem holds in a strictly convex, smooth normed plane $(\mathbb{M}^2, \|\cdot\|)$, then this plane is Euclidean.*

*Proof.* Consider $\Sigma = (\mathbb{M}^2, \|\cdot\|)$ as a Möbius plane. According to the Theorem of Smid and van der Waerden (see [90] and [11, § 5]), this plane is isomorphic to a Möbius plane $\Sigma' = \text{Mo}(\mathbb{F}, \mathbb{E})$, where $\mathbb{F}$ is a commutative field and $\mathbb{E}$ is a quadratic extension of $\mathbb{F}$. Denote by $\theta$ the corresponding homography from $\Sigma$ to $\Sigma'$. The plane $(\mathbb{M}^2, \|\cdot\|)$ is an (F)-plane, therefore $\Sigma'$ is also an (F)-plane. But Theorem 5 in [10] states that if $\Sigma'$ is an (F)-plane, then $\mathbb{E}$ is a separable extension of $\mathbb{F}$. If $\mathcal{C}$ is the unit circle of $(\mathbb{M}^2, \|\cdot\|)$, then let $\theta(\mathcal{C}) = \mathcal{C}'$. Since $\mathbb{E}$ is a separable extension of $\mathbb{F}$, there exists exactly one homography $\psi$ of $\Sigma'$ being an involution that fixes only the points of $\mathcal{C}'$; see [47], but also [11, § 4.7]. Therefore $\theta^{-1}\psi\theta$ is the inversion with respect to $\mathcal{C}$. Thus Theorem 3.4.5 implies that $(\mathbb{M}^2, \|\cdot\|)$ is Euclidean. $\square$

**Remark 3.4.4.** *Consider four pairs from the set of points $p_1, \ldots, p_8$ such that the points in every such pair are different and every point belongs to exactly one pair, e.g.,*

$$(p_1, p_2), (p_3, p_4), (p_5, p_6), (p_7, p_8).$$

## 3.5. Visibility in packing of balls

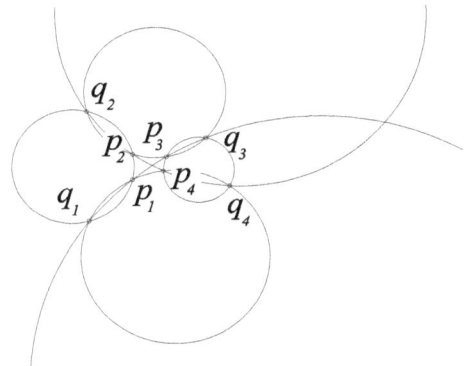

Figure 3.7: Bundle Theorem.

*These four pairs can be combined as pairs in six ways. Thus we obtain the quadruples*

$$(p_1, p_2, p_3, p_4), (p_1, p_2, p_5, p_6), (p_1, p_2, p_7, p_8),$$
$$(p_3, p_4, p_5, p_6), (p_3, p_4, p_7, p_8), (p_5, p_6, p_7, p_8). \tag{3.25}$$

*If five quadruples in (3.25) are concyclic, then such a configuration is called a* **Bundle con-figuration**. *The statement that* all six *quadruples in a Bundle configuration are concyclic (see Figure 3.7) is known as the* **Bundle theorem**. *If Miquel's theorem holds in a Möbius plane, then also the Bundle theorem holds there; see [90] and [11, § 5]. But the converse implication is not always true; see again [11, § 5]. In [89] the following theorem is given: if in a strictly convex, smooth normed plane* $(\mathbb{M}^2, \|\cdot\|)$ *the Bundle theorem holds, then* $(\mathbb{M}^2, \|\cdot\|)$ *is Euclidean. Thus we can derive another proof of our Theorem 3.4.6.*

## 3.5 Visibility in packing of balls

Let $K$ be a *convex body* in $\mathbb{M}^d$ and $p + K$ and $q + K$ be two disjoint translates of $K$. A translate $r + K$ is said to *lie between* $p + K$ *and* $q + K$ if $r + K$ overlaps none of $p + K$ and $q + K$ and there exist points $x \in (p + K)$ and $y \in (q + K)$ such that $[x, y] \cap (r + K) \neq \emptyset$. Assume that $r + K$ lies between $p + K$ and $q + K$. For $x \in (p + K)$ and $y \in (q + K)$, the segment $[x, y]$ is said to be *stably blocked* by $r + K$ if $[x, y] \cap \text{int}(r + K) \neq \emptyset$.

Let $\mathfrak{F}$ be a family of translates of $K$ packed between disjoint translates $p + K$ and $q + K$. Then $p + K$ and $q + K$ are called *visible* from each other in the packing $\{p + K, q + K\} \cup \mathfrak{F}$ if there exist points $x \in (p + K)$ and $y \in (q + K)$ such that the segment $[x, y]$ intersects no element of $\mathfrak{F}$. Otherwise $p + K$ and $q + K$ are said to be *concealed* from each other by $\mathfrak{F}$.

In this section we shall deal with translates of a centrally symmetric convex body $K$, and then we can assume that $K$ is the *unit ball* $\mathcal{B}$ (*unit disc* $\mathcal{D}$ in the planar case) of a normed space $(\mathbb{M}^d, \|\cdot\|)$.

The *concealment number* $\delta((\mathbb{M}^d, \|\cdot\|))$ is defined as the infimum of $\lambda > 0$ satisfying the following condition: for $\mathcal{B}(p,1)$ and $\mathcal{B}(q,1)$ being disjoint, the inequality $\|p-q\| > \lambda$ implies that $\mathcal{B}(p,1)$ and $\mathcal{B}(q,1)$ can be concealed from each other by packing translates of $\mathcal{B}$ between them. If $\|p-q\| \geq 4$, then $\mathcal{B}(p,1)$ and $\mathcal{B}(q,1)$ are concealed by $\mathcal{B}(\frac{p+q}{2},1)$, which implies that for any norm

$$\delta((\mathbb{M}^d, \|\cdot\|)) \leq 4.$$

It is also easy to check that $\delta(\mathbb{E}^2) = 2\sqrt{3}$. In [48] more consideration about $\delta(\mathbb{E}^2)$ can be found. It is our aim in this section to continue the investigations in [48] by proving several results on concealment numbers for two-dimensional normed spaces.

It should be also noticed that the visibility in crowds of translates of a convex body can be studied with the help of the corresponding visibility graph. For such an approach we refer to [49].

### 3.5.1 Special and very special triangles

We say that a triangle with vertices $p, q, r \in (\mathbb{M}^2, \|\cdot\|)$ is a *special triangle with base* $[p,q]$ if $\|p-r\| = \|q-r\| = 2$ and $\delta(r, \mathrm{L}(p,q)) = 1$. When we refer to $\mathcal{T}(p,r,q)$ as a special triangle we shall assume that $[p,q]$ is its base. If $\mathcal{T}(p,r,q)$ is special, then $\mathcal{T}(p, p+q-r, q)$ is also special but with different orientation. Any translate of a special triangle is a special triangle.

**Proposition 3.5.1.** *For any direction there exists a special triangle with base parallel to this direction. All special triangles with the same orientation and bases parallel to a given direction are corresponding to each other with respect to some translation.*

*Proof.* Let $G$ and $G'$ be two parallel lines such that $\delta(G, G') = 1$. Take $r \in G'$ and let $p$ and $q$ be the only two points where $G$ cuts $\mathcal{C}(r,2)$. Then $\mathcal{T}(p,r,q)$ is a special triangle. Similarly, another special triangle with the same base but with different orientation can be constructed by considering $G'$ in the other half-plane defined by $G$. The second part of the lemma follows from the construction of $\mathcal{T}(p,r,q)$. □

**Proposition 3.5.2.** *Let $\mathcal{T}(p,r,q)$ be a special triangle and let $t \in \mathrm{L}(p,q)$ be such that $\delta(r, \mathrm{L}(p,q)) = \|r-t\| = 1$. Then $r - t \dashv p - q$ and $t = \alpha p + (1-\alpha)q$, with $\frac{1}{4} \leq \alpha \leq \frac{3}{4}$.*

*Proof.* Assume that $t = \alpha p + (1-\alpha)q$. Then $\|r-t\| \leq \|r - (\mu p + (1-\mu)q)\|$ for every $\mu \in \mathbb{R}$. Taking $\mu = \alpha - \lambda$, we get that $\|r-t\| \leq \|r - t + \lambda(p-q)\|$ for every $\lambda \in \mathbb{R}$, i.e., $r - t \dashv p - q$.

Moreover, $1 = \|r-t\| = \|\alpha(r-p)+(1-\alpha)(r-q)\| \geq |\|\alpha(r-p)\|-\|(1-\alpha)(r-q)\|| = 2|\|\alpha|-|1-\alpha||$, which implies that $\frac{1}{4} \leq \alpha \leq \frac{3}{4}$. □

**Proposition 3.5.3.** *If $\mathcal{T}(p,r,q)$ is a special triangle, then $2 \leq \|p - q\| \leq 4$. Moreover, $\|p - q\| = 2$ if and only if the unit circle $\mathcal{C}$ is a parallelogram and $L(p,q)$ is parallel to some of the diagonals of $\mathcal{C}$.*

*Proof.* Let $\mathcal{T}(p,r,q)$ be a special triangle. Then $\|p-q\| \leq \|p-r\| + \|r-q\| = 4$. From Proposition 3.5.2 it follows that there exists $t \in [p,q]$ such that $\|r-t\| = 1$. This implies that $\|p-q\| = \|p-t\| + \|q-t\| \geq \|p-r\| - \|r-t\| + \|q-r\| - \|r-t\| = 2$. Suppose now that $\|p-q\| = 2$. Without loss of generality we can assume that $r = 0$. Let $t = \alpha p + (1-\alpha)q$ be as in Proposition 3.5.2. Then $2 = \|p\| = \|t + (1-\alpha)(p-q)\| \leq 1 + 2(1-\alpha)$, which implies $\alpha \leq \frac{1}{2}$. On the other hand, $2 = \|q\| = \|t + \alpha(q-p)\| \leq 1 + 2\alpha$, and then $\alpha = \frac{1}{2}$. Therefore the eight points $\frac{\pm(p+q)}{2}, \frac{\pm(p-q)}{2}, \frac{\pm p}{2}, \frac{\pm q}{2}$ belong to $\mathcal{C}$, which implies that $\mathcal{C}$ is the parallelogram with vertices $\frac{\pm p \pm q}{2}$. Conversely, assume that $\mathcal{C}$ is a parallelogram with vertices $\pm u, \pm v$, and let $\mathcal{T}(p,r,q)$ be a special triangle such that $\langle p,q \rangle$ is parallel to $\langle u, -u \rangle$. Then, either $\{p,q\} = \{r-u+v, r+u+v\}$ or $\{p,q\} = \{r-u-v, r+u-v\}$. In both cases, $\|p-q\| = \|2u\| = 2$. □

**Remark 3.5.1.** *Attaining the bound 4 in Proposition 3.5.3 does not fix the shape of $\mathcal{C}$. Consider in $\mathbb{M}^2$ the points $r = (1,0)$, $u = (\frac{1}{2}, 1)$, and $v = (-\frac{1}{2}, 1)$, and let $\|\cdot\|$ be any norm whose unit circle $\mathcal{C}$ contains $r$ and $[u,v]$, and $r \dashv u + v$. For $p = u + v$ and $q = -u - v$, the triangle $\mathcal{T}(p,r,q)$ is special and $\|p-q\| = 4$.*

Let $\mathcal{T}(p,r,q)$ be a special triangle, and $[\alpha_1, \alpha_2]$ be the largest interval such that $\|r - (\alpha p + (1-\alpha)q)\| = 1$ for $\alpha \in [\alpha_1, \alpha_2]$. From Proposition 3.5.2 it follows that $[\alpha_1, \alpha_2] \subset [\frac{1}{4}, \frac{3}{4}]$. Moreover, $\alpha_1 = \alpha_2$ if and only if the unit circle $\mathcal{C}$ does not contain a segment parallel to $L(p,q)$. We call $\mathcal{T}(p,r,q)$ a *very special triangle* if $\frac{1}{2} \in [\alpha_1, \alpha_2]$. In other words, a special triangle $\mathcal{T}(p,r,q)$ is very special if $L(r, \frac{p+q}{2}) \dashv L(p,q)$.

**Proposition 3.5.4.** *A normed plane $(\mathbb{M}^2, \|\cdot\|)$ is Euclidean if and only if every special triangle is a very special triangle.*

*Proof.* Obviously, in the Euclidean plane every special triangle is a very special triangle. Conversely, assume that every special triangle is a very special triangle. If $x, y \in \mathcal{C}$ are such that $\inf\{\|\alpha x + (1-\alpha)y\| : 0 \leq \alpha \leq 1\} = \frac{1}{2}$, then the triangle $\mathcal{T}(2x, 0, 2y)$ is special, and therefore very special, which gives that $\|x+y\| = 1$. Theorem 3.4.4 implies that the plane is Euclidean. □

## 3.5.2 The concealment number in the planar case

Let $\mathcal{C}$ be the unit circle of $(\mathbb{M}^2, \|\cdot\|)$ and let $p \in \mathcal{C}$. The *concealment number of the direction* $L(0, p)$, denoted by $\delta_p$, is defined as the infimum of $\mu > 2$ such that the unit disc $\mathcal{D}$ and its translate $\mathcal{D}(\mu p, 1)$ can be concealed from each other by packing translates of $\mathcal{D}$. Clearly, $\delta_p = \delta_{-p}$ and $\delta((\mathbb{M}^2, \|\cdot\|)) = \sup\{\delta_p : p \in \mathcal{C}\}$.

**Proposition 3.5.5.** *For any $p \in \mathcal{C}$, $2 \le \delta_p \le 4$. Moreover, $\delta_p = 2$ if and only if $\mathcal{C}$ is a parallelogram with $p$ as a vertex.*

*Proof.* It is obvious that $\delta_p \ge 2$. Since $\mathcal{D}$ and $\mathcal{D}(4p, 1)$ are concealed by $\mathcal{D}(2p, 1)$, we have that $\delta_p \le 4$.

Assume now that $\delta_p = 2$. Then for every $n \in \mathbb{N}$ there exists $2 < \mu_n \le 2 + \frac{1}{n}$ such that $\mathcal{D}$ and $\mathcal{D}(\mu_n p, 1)$ can be concealed. Therefore, for each $n$ there exists an $x_n$ such that $\mathcal{D}(x_n, 1)$ is between $\mathcal{D}$ and $\mathcal{D}(\mu_n p, 1)$, and $\mathcal{D}(x_n, 1) \cap [0, \mu_n p] \ne \emptyset$. Let $\alpha_n p \in \mathcal{D}(x_n, 1) \cap [0, \mu_n p]$. Then $1 \le \alpha_n \le \mu_n - 1$, $\|x_n\| \ge 2$, $\|x_n - \mu_n p\| \ge 2$, and $\|x_n - \alpha_n p\| \le 1$, which implies that

$$\mu_n = \|\alpha_n p\| + \|\mu_n p - \alpha_n p\| \ge \|x_n\| - \|x_n - \alpha_n p\| + \|x_n - \mu_n p\| - \|x_n - \alpha_n p\| \ge 2.$$

Since $\mu_n \to 2$ as $n \to \infty$, it follows that $\alpha_n \to 1$ and $x_n \to x$ with $\|x\| = \|x - 2p\| = 2$ and $\|x - p\| = 1$. We thus get that $\mathcal{C}$ is the parallelogram with vertices $\pm p, \pm(x - p)$.

Conversely, assume that $\mathcal{C}$ is a parallelogram with vertices $\pm p, \pm q$. Then, for any $\mu > 2$, $\{\mathcal{D}(\frac{\mu}{2}p + q, 1), \mathcal{D}(\frac{\mu}{2}p - q, 1)\}$ conceals $\mathcal{D}$ and $\mathcal{D}(\mu p, 1)$, which implies that $\delta_p = 2$. □

**Remark 3.5.2.** *It is easy to give examples showing that the identity $\delta_p = 4$ does not determine the shape of $\mathcal{C}$.*

The next theorem relates a special triangle with the concealment number of the direction of its base.

**Theorem 3.5.1.** *Let $p, q \in (\mathbb{M}^2, \|\cdot\|)$.*

(i) *If $[p, q]$ is the base of a special triangle, then $\delta_{\frac{p-q}{\|p-q\|}} \ge \|p - q\|$.*

(ii) *If $[p, q]$ is the base of a very special triangle, then $\delta_{\frac{p-q}{\|p-q\|}} = \|p - q\|$.*

(iii) *If $\|p - q\| < 4$ and $[p, q]$ is the base of a special triangle that is not very special, then $\delta_{\frac{p-q}{\|p-q\|}} > \|p - q\|$.*

*Proof.* (i) Let $p, q \in (\mathbb{M}^2, \|\cdot\|)$ be such that $[p, q]$ is the base of a special triangle, and let $\bar{p} = \frac{p-q}{\|p-q\|}$. Then $[0, \|p - q\|\bar{p}]$ is also the base of a special triangle. To prove that $\delta_{\bar{p}} \ge \|p - q\|$ we shall assume, on the contrary, that there exists $2 < \mu < \|p - q\|$ such that $\mathcal{D}$ and $\mathcal{D}(\mu\bar{p}, 1)$ can be concealed by a family of translates of $\mathcal{D}$. We shall get a contradiction.

## 3.5. Visibility in packing of balls

Let $\mathcal{D}(x,1)$ be such that neither $\mathcal{D}$ nor $\mathcal{D}(\mu\bar{p},1)$ are overlapped by it, and $\mathcal{D}(x,1) \cap [0,\mu\bar{p}] \neq \emptyset$. Then $\|x\| \geq 2$ and $\|x - \mu\bar{p}\| \geq 2$. Moreover, $x \notin [0,\mu\bar{p}]$, because in the other case $\|p-q\| > \mu \geq 4$, contradicting Proposition 3.5.3. Let $x_1 \in \mathcal{D}(x,1) \cap [0,\mu\bar{p}]$. We can assume that $x - x_1 \dashv \bar{p}$, since any $x_1' \in L(0,\bar{p})$ such that $x - x_1' \dashv \bar{p}$ satisfies $\|x - x_1'\| \leq \|x - x_1\|$, and then $x_1' \in \mathcal{D}(x,1)$. Let $p_1 = \frac{x-x_1}{\|x-x_1\|}$, $q_1 = p_1 + \mu\bar{p}$, and $q_1' = p_1 + \|p-q\|\bar{p}$. Let $y \in L(p_1,q_1)$ be such that $\mathcal{T}(0,y,p-q)$ is a special triangle. Now we distinguish two situations:

(a) The line $L(0,\bar{p})$ supports $\mathcal{D}(x,1)$, i.e., $x \in L(p_1,q_1)$, and $x$ is strictly between $p_1$ and $q_1$. Thus $x = p_1 + \lambda\bar{p}$, with $0 < \lambda < \mu$. Since $\|y\| = 2 \leq \|x\|$, it follows from Lemma 1.2.3 and the convexity of $\mathcal{D}$ that $y$ is between $p_1$ and $x$. Again it follows from Lemma 1.2.3 that

$$2 \leq \|x - \mu\bar{p}\| \leq \|x - (p-q)\| = \|x - \|p-q\|\bar{p}\| \leq \|y - \|p-q\|\bar{p}\| = 2,$$

and then $\|x - \mu\bar{p}\| = \|x - \|p-q\|\bar{p}\| = 2$. But from the identity

$$(\|p-q\| - \lambda)(x - \mu\bar{p}) = (\mu - \lambda)(x - \|p-q\|\bar{p}) + (\|p-q\| - \mu)p_1$$

it follows that $2(\|p-q\| - \lambda) \leq 2(\mu - \lambda) + \|p-q\| - \mu$, which gives $\|p-q\| \leq \mu$, against the hypothesis.

(b) The segment $[0,\mu\bar{p}]$ is stably blocked by $\mathcal{D}(x,1)$. Then $x$ lies in the interior of the parallelogram with vertices $0$, $p_1$, $q_1$, and $\mu\bar{p}$. Therefore $x$ is also an interior point of the parallelogram with vertices $0$, $p_1$, $q_1'$, and $p-q$. Assume that $x \in \operatorname{conv}\{0,y,p-q\}$. Then, from Lemma 1.2.2, we get

$$4 \leq \|x\| + \|x - \mu\bar{p}\| \leq \|x\| + \|x - (p-q)\| \leq \|y\| + \|y - (p-q)\| = 4.$$

Therefore $\|x\| = \|x - \mu\bar{p}\| = \|x - (p-q)\| = 2$, which implies that the segment $[0, p-q]$ belongs to $\mathcal{C}(x,2)$, contradicting that $[0,\mu\bar{p}]$ is stably blocked by $\mathcal{D}(x,1)$. Therefore, either $x \in \operatorname{conv}\{0,y,p_1\}$ or $x \in \operatorname{conv}\{p-q,y,q_1'\}$. The first situation is impossible because if $x' = \mathrm{R}_x^+(0) \cap L(p_1,q_1)$, then $x' \in [p_1,y]$, which implies the absurdity $2 \leq \|x\| < \|x'\| \leq \|y\| = 2$. The second situation is also impossible because taking $x'' = \mathrm{R}_x^+(p-q) \cap L(p_1,q_1)$, we have that $x'' \in [y,q_1']$, and then $2 \leq \|x - \mu\bar{p}\| \leq \|x - (p-q)\| < \|x'' - (p-q)\| \leq \|y - (p-q)\| = 2$, which is absurd, too.

(ii) Assume that $[p,q]$ is the base of a very special triangle, and let $x$ be such that $\mathcal{T}(0,x,p-q)$ is a very special triangle. Then $\mathcal{T}(0,p-q-x,p-q)$ is also a very special triangle; therefore $\frac{1}{2}(p-q) \in \mathcal{C}(x,1) \cap \mathcal{C}(p-q-x,1)$, and $L(0,p-q)$ supports both circles at this point. This implies that $[x,p-q-x] \subset \mathcal{D}(x,1) \cup \mathcal{D}(p-q-x,1)$. Since for any $u \in \mathcal{D}$, $v \in \mathcal{D}(p-q,1)$ the segment $[u,v]$ intersects the segment $[x,p-q-x]$, we have that $\{\mathcal{D}(x,1), \mathcal{D}(p-q-x,1)\}$ conceal $\mathcal{D}$ and $\mathcal{D}(p-q,1)$. Therefore, $\|p-q\| \geq \delta_{\frac{p-q}{\|p-q\|}}$. Part (i) completes the proof.

(iii) Assume that $\mathcal{T}(p,r,q)$ is a special triangle which is not very special such that $\delta_{\frac{p-q}{\|p-q\|}} = \|p-q\|$. We shall see that $\|p-q\| = 4$. We have that the line $\mathrm{L}(r, \frac{1}{2}(p+q))$ is not normal to $\mathrm{L}(p,q)$. Let $L$ be a line through $\frac{1}{2}(p+q)$ with $\mathrm{L}(r, \frac{1}{2}(p+q)) \dashv L$. Since $L \neq \mathrm{L}(p,q)$ and $L$ does not pass through $r$, then $L$ intersects either the segment $[r,q]$ or the segment $[r,p]$. Assume, without loss of generality, that $L$ intersects $[r,q]$ in a point $w$ which is different to $r$ and $q$. Since $d(r,L) > 1$, we have $\|w-q\| < 1$. Let $w' = p+q-w$. Then $w' \in L \cap \mathcal{D}(p,1)$. For $n \in \mathbb{N}$, let $q_n = q + \frac{1}{n}(q-p)$. Then $\|p-q_n\| = (1+\frac{1}{n})\|p-q\| > \|p-q\|$. Since $\delta_{\frac{p-q}{\|p-q\|}} = \|p-q\|$, there exists a point between $q$ and $q_n$, still denoted by $q_n$, such that $\mathcal{D}(p,1)$ and $\mathcal{D}(q_n,1)$ can be concealed. Moreover, since $L$ meets the interior of $\mathcal{D}(q,1)$, for sufficiently large $n$ also $L$ meets $\mathcal{D}(q_n,1)$. Hence there exists an $x_n$ such that $\mathcal{D}(x_n,1)$ is between $\mathcal{D}(p,1)$ and $\mathcal{D}(q_n,1)$, and $\mathcal{D}(x_n,1)$ intersects $L$ in a point $t_n$. Thus $\|p-x_n\| \geq 2$, $\|q_n-x_n\| \geq 2$, and $\|t_n-x_n\| \leq 1$. Letting $n$ tend to infinity, we have that (for a subsequence, if necessary) $x_n \to x$ and $t_n \to t$, such that $\|p-x\| \geq 2$, $\|q-x\| \geq 2$, $\|t-x\| \leq 1$, and $t \in [w',w] \subset L$, which implies that $d(x,L) \leq 1$, and then $x \neq r$. Without loss of generality we can assume that $x$ is in the same half-plane defined by $L$ as $r$ is. In the other case we can consider $r' = q+p-r$ instead of $r$. Let $L' = (r - \frac{1}{2}(p+q)) + L$. Thus $x$ is between the lines $L$ and $L'$. Let $L_p$ and $L_q$ be lines through $p$ and $q$, respectively, parallel to $\mathrm{L}(r, \frac{1}{2}(p+q))$. We shall see that $x$ is between $L_p$ and $L_q$. On the contrary, assume first that $L_q$ is between $r$ and $x$. Let $\bar{L}$ be the line parallel to $L$ that supports $\mathcal{D}(q,1)$ at a point $\bar{q} \in L_q$ between $L$ and $L'$. Then $L \neq \bar{L}$, because $w \in L$ and $\|w-q\| < 1$. If $x$ is between $L$ and $\bar{L}$, then $[x,t]$ cuts $[q,\bar{q}]$ at an interior point of $\mathcal{D}(q,1)$, which is contradictory. Then assume that $x$ is between $\bar{L}$ and $L'$. Since $d(q,L') \leq \|q-r\| = 2$ and $d(q,\bar{L}) = 1$, we have that $d(\bar{L},L') \leq 1$. Let $L_x$ be the line through $x$ parallel to $L_q$, and let $\{\bar{x}\} = \bar{L} \cap L_x$. Then $\bar{x} \in \mathcal{D}(x,1)$, and $[\bar{x},t]$ cuts the interior of $\mathcal{D}(q,1)$, which is again contradictory. On the other hand, if $L_p$ is between $x$ and $r$, then $[x,t]$ trivially intersects $\mathcal{D}(p,1)$. Therefore $x$ is between $L_p$ and $L_q$ and between $L$ and $L'$. But then, from Lemma 1.2.3 it follows that $x \in \mathrm{conv}\{p,q,r\}$. Let $\{z\} = \mathrm{L}(p,q) \cap \mathrm{L}(r,x)$, and let $0 \leq \mu < 1$ be such that $x = \mu r + (1-\mu)z$. Consider the convex function $f(\lambda) = \|p-z+\lambda(z-r)\| + \|q-z+\lambda(z-r)\|$. Then $f(1) = \|p-r\| + \|q-r\| = 4$ and $f(\mu) = \|p-x\| + \|q-x\| \geq 4$, which implies that $4 \leq f(0) = \|p-z\| + \|q-z\| = \|p-q\|$. Finally, Proposition 3.5.3 gives $\|p-q\| = 4$. $\square$

**Remark 3.5.3.** *In view of Theorem 3.5.1(iii), Figure 3.8 gives an example of a normed plane* $(\mathbb{M}^2, \|\cdot\|)$ *and a special triangle* $\mathcal{T}(p,r,q)$ *with* $\|p-q\| = 4$ *that is not very special with* $\delta_{\frac{p-q}{\|p-q\|}} = \|p-q\|$.

We will give a geometric characterization of all the directions for which the concealment number is precisely determined. For that reason we need the following

3.5. VISIBILITY IN PACKING OF BALLS                                                                 81

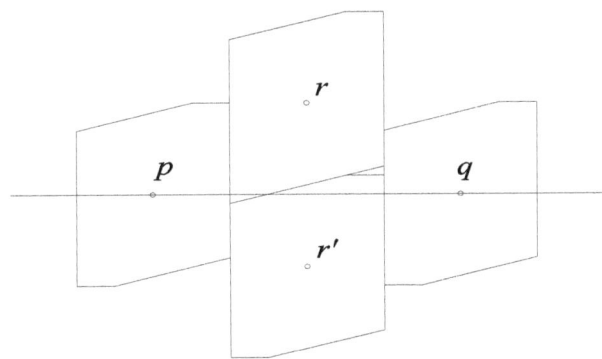

Figure 3.8: A special triangle $\mathcal{T}(p,r,q)$ with $\|p-q\| = 4$ that is not very special.

**Lemma 3.5.1.** *Assume that $p_1$, $p_2$, $q_1$, $q_2$ are four distinct points on the unit circle $\mathcal{C}$ of $(\mathbb{M}^2, \|\cdot\|)$ such that $p_1 \prec q_1$, $p_2 \prec q_2$, $p_1 \prec p_2$, and $\frac{p_1+q_1}{2} = \frac{p_2+q_2}{2} = x$ with $0 < \|x\| < 1$. Then the points $p_1$, $p_2$, $-q_1$, $-q_2$ are aligned and the segment containing them belongs to $\mathcal{C}$.*

*Proof.* From the hypothesis it follows that the four points are in one of the following locations: (a) $p_1 \prec q_1 \prec p_2 \prec q_2$; (b) $p_1 \prec p_2 \prec q_2 \prec q_1$; (c) $p_1 \prec p_2 \prec q_1 \prec q_2$. Case (a) is impossible because the segments $[p_1, q_1]$ and $[p_2, q_2]$ have the point $x$ in common. The same reason and the convexity of the unit disc imply that in case (b) the four points are aligned, which implies that the segment $[p_1, q_1]$ belongs to $\mathcal{C}$ and then $\|x\| = 1$, against the hypothesis. Finally, assume that (c) occurs. Since $\|x\| > 0$, we have that $p_1 \neq -q_1$ and $p_2 \neq -q_2$. Then $\{p_1, q_1, -p_1, -q_1\}$ and $\{p_2, q_2, -p_2, -q_2\}$ define two parallelograms with vertices in $\mathcal{C}$ such that the sides $[p_1, -q_1]$ and $[p_2, -q_2]$ are parallel and of equal length, and the sides $[p_1, q_1]$ and $[p_2, q_2]$ meet at the midpoint $x$. Moreover, $\{p_1, p_2, q_1, q_2\}$ define another parallelogram with the sides $[p_1, p_2]$ and $[q_1, q_2]$ parallel. From the convexity of the unit disc it follows that no vertex of any of these parallelograms can be in the interior of any other parallelogram. Therefore, either $[p_1, q_1]$ is parallel to $[p_2, q_2]$, which contradicts $\|x\| < 1$, or the points $p_1, p_2, -q_1, -q_2$ are aligned, and then the segment that contains them belongs to $\mathcal{C}$. □

**Theorem 3.5.2.** *The base $[p, q]$ of a special triangle in a normed plane $(\mathbb{M}^2, \|\cdot\|)$ with $\|p-q\| < 4$ is the base of a very special triangle if and only if at least one of the chords of the unit circle $\mathcal{C}$ having $\frac{1}{4}(q-p)$ as midpoint is normal to $\mathrm{L}(p,q)$.*

*Proof.* Let $\mathcal{T}(p,r,q)$ be a special triangle with $\|p-q\| < 4$, and let $u = \frac{1}{2}(r-p)$, $v = \frac{1}{2}(q-r)$. Then $u, v \in \mathcal{C}$, $\frac{1}{2}(u+v) = \frac{1}{4}(q-p)$ and $u-v = r - \frac{1}{2}(p+q)$. Assume now that $\mathcal{T}(p,r,q)$

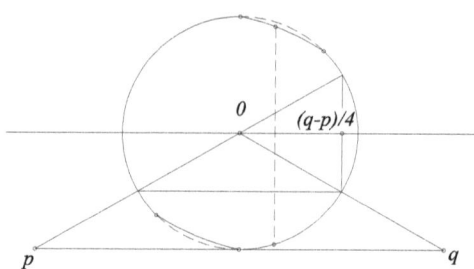

Figure 3.9: Not always the condition $\delta_{\frac{p-q}{\|p-q\|}} = \|p-q\|$ implies the existence of reflections in the line L$(p,q)$.

is very special. Then $u - v \dashv p - q$, and therefore the chord of $\mathcal{C}$ defined by $u$ and $v$ has $\frac{1}{4}(q - p)$ as midpoint and is orthogonal to L$(p,q)$. Conversely, assume that $u', v' \in \mathcal{C}$ are such that $\frac{1}{2}(u' + v') = \frac{1}{4}(q - p)$ and $u' - v' \dashv p - q$. If $\{u', v'\} = \{u, v\}$, then $r - \frac{1}{2}(p + q) \dashv p - q$, and $\mathcal{T}(p, r, q)$ is very special. On the other hand, if $\{u', v'\} \neq \{u, v\}$, then $u, v, u', v'$ are four different (recall that $\|p - q\| < 4$) points in $\mathcal{C}$ such that the midpoints of the chords $[u, v]$ and $[u', v']$ coincide. Let $\prec$ be an orientation of the plane such that $u \prec v$. Without loss of generality we can assume that $u' \prec v'$. If $u \prec u'$, then from Lemma 3.5.1 it follows that $u, u', -v$, and $-v'$ are in a segment contained in $\mathcal{C}$. This segment is parallel to $u + v$, and therefore it is also parallel to $q - p$. Moreover, the point $\frac{1}{2}(u - v)$ belongs to this segment and then $u - v \dashv q - p$, which implies that $\mathcal{T}(p, r, q)$ is very special. If $u' \prec u$, we get the same result. $\square$

For strictly convex normed planes, Busemann and Kelly defined reflections in a line as isometries having this line as line of fixed points; see [27, p. 127]. They also proved that a reflection in every line exists if and only if the plane is Euclidean; cf. [27, p. 140, Theorem 25.3]. But the reflection in a line $G$ exists if and only if for every circle with center on $G$ the following holds: there exists a chord $C$ of that circle such that all chords parallel to $C$ are bisected by $G$; see [27, p. 140, Theorem 25.2]. Note that if one circle with center on $G$ has this property, then all such circles have it. If a line $G$ admits a reflection $\varphi$, then for every point $x \notin G$ we have L$(x, \varphi(x)) \dashv G$, and the midpoint of $[x, \varphi(x)]$ lies on $G$; see [27, p. 128, Theorem 23.4*]. Thus Theorem 3.5.2 and Theorem 3.5.1(ii) imply

**Corollary 3.5.1.** *Let* $p, q \in (\mathbb{M}^2, \|\cdot\|)$, *and* $[p, q]$ *be the base of a special triangle with* $\|p-q\| < 4$. *If the line* L$(p, q)$ *admits a reflection (i.e.,* $(\mathbb{M}^2, \|\cdot\|)$ *is strictly convex), then*

$$\delta_{\frac{p-q}{\|p-q\|}} = \|p - q\|.$$

## 3.5. VISIBILITY IN PACKING OF BALLS

**Remark 3.5.4.** *The example in Figure 3.5.2 shows that the condition $\delta_{\frac{p-q}{\|p-q\|}} = \|p-q\|$ not always implies the existence of reflections in the line $\mathrm{L}(p,q)$ .*

**Remark 3.5.5.** *A different approach to reflections in normed planes can be found in [67] and [68]. In contrast to the approach of Busemann and Kelly the reflections in lines there are defined as affine transformations that are nor necessarily isometries.*

# Bibliography

[1] ALONSO, J., MARTINI, H., MUSTAFAEV, Z.: On orthogonal chords in normed planes, *The Rocky Mountain J. Math.* **41** (2011), 23-36.

[2] ALONSO, J., MARTINI, H., SPIROVA, M.: Visibility in crowds of translates of a centrally symmetric body, *European J. Combin.* **31** (2010), 710-719.

[3] ALONSO, J., SPIROVA, M.: Characterizations of different classes of convex bodies via orthogonality, *Bull. Belg. Math. Soc. Simon Stevin*, 16 pp., to appear.

[4] ALVAREZ, J. C.: Some problems in Finsler geometry, The Finsler Geometry Newsletter, http://gauss.math.ucl.ac.be/~fweb/intro/intro.html (2000), 1-30.

[5] ÁLVARES PAIVA, J. C., THOMPSON, A. C.: On the perimeter and area of the unit disc, *Amer. Math. Monthly* **112** (2005), 141-154.

[6] ASPLUND, E., GRÜNBAUM, B.: On the geometry of Minkowski planes, *Enseign. Math.* **6** (1960), 299-306.

[7] AVERKOV, G.: On cross-section measures in Minkowski spaces, *Extracta Math.* **18** (2003), 201–208.

[8] BANASIAK, J.: Some contributions to the geometry of normed linear spaces, *Math. Nachr.* **139** (1988), 175-184.

[9] BENÍTEZ, C., YÁÑEZ, D.: Middle points, medians and inner products, *Proc. Amer. Math. Soc.* **135** (2007), 1725-1734.

[10] BENZ, W.: Beziehungen zwischen Orthogonalitäts- und Anordnungseigenschaften in Kreisebenen, *Math. Ann.* **134** (1957/58), 385-402.

[11] BENZ, W.: Über Möbiusebenen. Ein Bericht, *Jahresber. Deutsch. Math.-Verein.* **63** (1960), 1-27.

[12] BERGER, M.: *Geometry I*, Springer - Verlag, Berlin ect., 1987.

[13] BEZDEK, A., KUPERBERG, W.: Circle coverings with a margin, *Period. Math. Hungar.* **34** (1997), 3-16.

[14] BEZDEK, K., CONNELLY, R.: Covering circles by translates of a convex set, *Amer. Math. Monthly* **96** (1989), 789-806.

[15] BEZDEK, K., CONNELLY, R.: Pushing disks apart, *J. Reine Angew. Math.*, **553** (2003), 221–236.

[16] BEZDEK, K., CONNELY, R., CSIKÓS, B.: On the perimeter of the intersection of congruent disks, *Beiträge Algebra Geom.* **47** (2006), 53-62.

[17] BEZDEK, K., LÁNGI, Z., NASZÓDI, M., PAPEZ, P.: Ball-polyhedra, *Discrete Comput. Geom.* **38** (2007), 201-230.

[18] BEZDEK, K., NASZÓDI, M.: Rigidity of ball-polyhedra in Euclidean 3-space, *European J. Combin.* **27** (2006), 255–268.

[19] BIRKHOFF, G.: Orthogonality in linear metric spaces, *Duke Math. J.* **1** (1935), 169-172.

[20] BIRMAN, G., NOMIZU, K.: Trigonometry in Lorentzian geometry, *Amer. Math. Monthly* **91** (1984), 543-549.

[21] BLASCHKE, W.: Konvexe Bereiche gegebener konstanter Breite und kleinsten Inhalts, *Math. Annalen* **76** (1915), 504-513.

[22] BLUMENTHAL, L. M.: *Theory and Applications of Distance Geometry*, Second ed., Chelsea Publishing Co., New York, 1970.

[23] BOLTYANSKI, V., MARTINI, H., SOLTAN, P. S.: *Excursions into Combinatorial Geometry*, Springer, Berlin - Heidelberg, 1996.

[24] BÖRÖCZKY JR., K.: *Finite Packing and Covering*, Cambridge Tracts in Mathematics, Vol. 154, Cambridge University Press, 2004.

[25] BUSEMANN, H.: The isoperimetric problem in the Minkowski plane, *Amer. J. Math.* **69** (1947), 863-871.

[26] BUSEMANN, H.: *The Geometry of Geodesics*, Academic Press Inc., New York, N. Y., 1955.

[27] BUSEMANN, H., KELLY, P.: *Projective Geometry and Projective Metrics*, Academic Press Inc., New York, 1953.

[28] CAPOYLEAS, V.: On the area of the intersection of disks in the plane, *Comput. Geom.* **6** (1996), 393-396.

[29] CASPANI, L.: *Diametrically maximal and constant width sets in banach spaces*, PhD Thesis, Università Degli Studi di Milano, 2010.

[30] CHAKERIAN, G. D.: Sets of constant width, *Pacific J. Math.* **19** (1966), 13-21.

[31] CHAKERIAN, G. D.: Intersection and covering properties of convex sets, *Amer. Math. Monthly* **76** (1969), 753-766.

[32] CHAKERIAN, G. D., GROEMER, H.: Convex bodies of constant width, In: *Convexity and its applications*, Eds. P. M. Gruber and J. M. Wills, Birkhäuser, Basel, 1983, 49–96.

[33] DAY, M. M.: Some characterizations of inner product spaces, *Trans. Amer. Math. Soc.* **62** (1947), 320-337.

[34] DOYLE, P. G., LAGARIAS, J., RANDALL, D.: Self-packing of centrally symmetric bodies in $\mathbb{R}^2$, *Discrete Comp. Geometry* **8** (1992), 171-189.

[35] EGGLESTON, H. G.: *Convexity*, Cambridge Univ. Press, Cambridge, 1958.

[36] EGGLESTON, H. G.: Sets of constant width in finite dimensional Banach spaces, *Israel J. Math.* **3** (1965), 163–172.

[37] FEJES TÓTH, G.: Packing and covering. In: *Handbook of Discrete and Computational Geometry*, Chapter 2. Eds. J. E. Goodman and J. O'Rourke, CRC Press, Ser. Discrete Math. Appl., 1997.

[38] FINSLER, P.: *Über Kurven und Flächen in allgemeinen Räumen*, Ph.D. thesis, Göttingen, 1918.

[39] GARDNER, R. J.: *Geometric Tomography*, Second Ed., Encyclopedia of Mathematics and its Applications, No. 58, Cambridge University Press, Cambridge, 2006.

[40] GROH, H.: Flat Moebius planes, *Geom. Dedicata* **1** (1972), 65-84.

[41] GRÜNBAUM, B.: Borsuk's partition conjecture in Minkowski planes, *Bull. Res. Council Israel Sect.* F **7F** (1957/1958), 25-30.

[42] GRÜNBAUM, B.: On some covering and intersection properties in Minkowski spaces, *Pacific J. Math.* **9** (1959), 487-494.

[43] GRÜNBAUM, B.: On a conjecture of H. Hadwiger, *Pacific J. Math.* **11** (1961), 215-219.

[44] HADWIGER, H.: Über Treffanzahlen bei translationsgleichen Eikörpern, *Arch. Math.* **8** (1957), 212-213.

[45] HILBERT, D.: Mathematical problems, *Bull. Amer. Math. Soc.* **37** (2000), 407–436.

[46] HOLUB, J. R.: Rotundity, orthogonality, and characterizations of inner product spaces, *Bull. Amer. Math. Soc.* **81** (1975), 1087-1089.

[47] HOFFMAN, A. J.: Chains in the projective line, *Duke Math. J.* **18** (1951), 827-830.

[48] HOSONO, K., MAEHARA, H., MATSUDA, K.: A pair in a crowd of unit balls, *European J. Combin.* **22** (2001), 1083-1092.

[49] HOSONO, K., MEIJER, H., RAPPAPORT, D.: On the visibility graph of convex translates, *Discrete Appl. Math.* **113** (2-3) (2001), 195-210.

[50] LACHAND-ROBERT, T., OUDET, É.: Bodies of constant width in arbitrary dimension, *Math. Nachr.* **280** (2007), 740–750.

[51] JAMES, R. C.: Inner products in normed linear spaces, *Bull. Amer. Math. Soc.* **53** (1947), 559-566.

[52] JAMES, R. C.: Orthogonality and linear functionals in normed linear spaces, *Trans. Amer. Math. Soc.* **61** (1947), 265-292.

[53] LASSAK, M.: Covering a plane convex body by four homothetical copies with the smallest positive ratio, *Geom. Dedicata* **21** (1986), 157-167.

[54] LASSAK, M.: Covering plane convex bodies with smaller homothetical copies, *Colloquia Mathematica Societatis János Bolyai* **48** (Intuitive Geometry, Siófok, 1985) North-Holland Publ. Co., Amsterdam, 1987, 331-337.

[55] LEBESQUE, H.: Sur le problème des isopérimètres et sur les domaines de largeur constante, *Bull. Soc. Math. France,* C. R. (1914), 72-76.

[56] LEVI, F. W.: Ein geometrisches Überdeckungsproblem, *Arch. Math.* **5** (1954), 476-478.

[57] KELLY, P. J.: On Minkowski bodies of constant width, *Bull. Amer. Math. Soc.* **55** (1949), 1147–1150.

[58] KUPITZ, Y. S., MARTINI, H., PERLES, M. A.: Finite sets in $\mathbb{R}^d$ with many diameters - a survey, In: *Proceedings of the International Conference on Mathematics and*

*Applications* (ICMA-MU 2005, Bangkok), Mahidol University Press, Bangkok, 2005, 91-112.

[59] MAKAI, E., JR., MARTINI, H.: A new characterizations of convex plates of constant width, *Geom. Dedicata* **34** (1990), 199-209.

[60] MARTINI, H., SPIROVA, M.: The Feuerbach circle and orthocentricity in normed planes, *Enseign. Math.* **53** (2007), 237-258.

[61] MARTINI, H., SPIROVA, M.: Recent results in Minkowski geometry. In: Proc. Internat. Conf. Math. Appl. (Mahidol University of Bangkok, Thailand, 2007), Mahidol University Press, Bangkok, 2007, pp. 45-83. (It also appeared in a special volume of the *East-West J. Math.* (2007), 59-101.)

[62] MARTINI, H., SPIROVA, M.: Clifford's chain of theorems in strictly convex Minkowski planes, *Publ. Math. Debrecen* **72** (2008), 371-383.

[63] MARTINI, H., SPIROVA, M.: Reflections in stricly convex Minkowski planes, *Aequationes Math.* **78** (2009), 71-85.

[64] MARTINI, H., SPIROVA, M.: Covering discs in Minkowski planes, *Canad. Math. Bull.* **52** (2009), 424-434.

[65] MARTINI, H., SPIROVA, M.: On the circular hull property in normed planes, *Acta Math. Hungar.* **125** (2009), 275-285.

[66] MARTINI, H., SPIROVA, M.: On regular 4-coverings and their applications for lattice coverings in normed planes, *Discrete Math.* **309** (2009), 5158-5168.

[67] MARTINI, H., SPIROVA, M.: Reflections in strictly convex Minkowski planes, *Aequationes Math.* **78** (2009), 71-85.

[68] MARTINI, H., SPIROVA, M., STRAMBACH, K.: Group-theoretical conditions in strictly convex Minkowski planes, 16 pp., submitted.

[69] MARTINI, H., SPIROVA, M., SWANEPOEL, K.: Geometry where directions matters - or does it? *Math. Intelligencer* **33**, no. 4 (2011), to appear.

[70] MARTINI, H., SWANEPOEL, K. J.: The geometry of Minkowski spaces - a survey. Part II, *Expositiones Math.* **22** (2004), 93-144.

[71] MARTINI, H., SWANEPOEL, K. J.: Antinorms and Radon curves, *Aequationes Math.* **72** (2006), 110-138.

[72] MARTINI, H., SWANEPOEL, K. J., WEISS, G.: The geometry of Minkowski spaces - a survey. Part I, *Expositiones Math.* **19** (2001), 97-142.

[73] MEISSNER, F.: Über Punktmengen konstanter Breite, *Vierteljahresschr. naturforsch. ges. Zürich* **56** (1911), 42-50.

[74] MENGER, K.: *Untersuchungen über allgemeine Metrik*, Math. Ann. **100** (1928), 75-163.

[75] MINKOWSKI, H.: *Sur les propriétés des nombres entiers qui sont dérivées de l'intuition de l'espace*, Nouvelles Annales de Mathematiques, 3e série **15** (1896), Also in Gesammelte Abhandlungen, 1. Band, XII, pp. 271-277.

[76] MORENO, J. P.: Porosity and unique completion in strictly convex spaces *Math. Z.* **267** (2011), 173-184.

[77] MORENO, J. P., SCHNEIDER, R.: Continuity properties of the ball hull mapping, *Nonlinear Anal.* **66** (2007), 914-925.

[78] MORENO, J. P., SCHNEIDER, R.: Intersection properties of polyhedral norms, *Adv. Geom.* **7** (2007), 391-402.

[79] MORENO, J. P., SCHNEIDER, R.: Local Lipschitz continuity of the diametrical completion mapping, submitted.

[80] OHMANN, D.: Extremalprobleme für konvexe Bereiche der euklidischen Ebene, *Math. Z.* **55** (1952), 346-352.

[81] PACH, J., AGARWAL, P. K.: *Combinatorial Geometry*, John Wiley and Sons, 1995.

[82] PERLES, M. A., MARTINI, H., KUPITZ, Y. S.: Ball polytopes and the Vászonyi problem, *Acta Math. Hungar.* **126** (2019), 99–163.

[83] RADON, J.: Über eine besondere Art ebener konvexer Kurven, *Ber. Verh. Sächs. Akad. Leipzig* **68** (1916), 123–128.

[84] RIEMANN, B.: *Über die Hypothesen, welche der Geometrie zu Grunde liegen*, Abh. Königlichen Gesellschaft Wiss. Göttingen **13** (1868).

[85] RUND, H.: *The Differential Geometry of Finsler Spaces*, Die Grundlehren der Mathematischen Wissenschaften, Bd. 101, Springer-Verlag, Berlin, 1959.

[86] SACHS, H.: *Ebene isotrope Geometrie*, Vieweg, Braunschweig-Wiesbaden, 1987.

[87] SALLEE, G. T.: The maximal set of constant width in a lattice, *Pacific J. Math.* **28** (1969), 669–674.

[88] SALLEE, G. T.: Maximal areas of Reuleaux polygons, *Canad. Math. Bull.*, **13** (1970), 175–179.

[89] SANTOS, F.: Inscribing a symmetric body in an ellipse *Inform. Process. Lett.* **59** (1996), 175-178.

[90] SMID, L. J., WAERDEN, B. L. VAN DER: Eine Axiomatik der Kreisgeometrie und der Laguerregeometrie, *Math. Ann.* **110** (1935), 753-776.

[91] SCHÄFFER, J. J.: Inner diameter, perimeter, and girth of spheres, *Math. Ann.* **173** (1967), 59-79; addendum **173** (1967), 79-82.

[92] SOLTAN, V.: Affine diameters of convex bodies - a survey, *Expositiones Math.* **23** (2005), 47–63.

[93] SPIROVA, M.: On a theorem of G. D. Chakerian, *Contrib. Discrete Math.* **5** (2010), 107–118.

[94] SPIROVA, M.: On Miquel's theorem and inversion in normed planes, *Monatsh. Math.*, **161** (2010), 335–345.

[95] SPIROVA, M.: Circle configurations in strictly convex normed planes, *Adv. Geom.*, **10** (2010), 361–346.

[96] STILES, W. J.: On inversions in normed linear spaces, *Proc. Amer. Math. Soc.* **20** (1969), 505-508.

[97] STRAMBACH, K.: Über sphärische Möbiusebenen, *Arch. Math.* **18** (1967), 208-211.

[98] STRAMBACH, K.: Sphärische Kreisebenen, *Math. Z.* **113** (1970), 266-292.

[99] THOMPSON, A. C.: *Minkowski Geometry*, Encyclopedia of Mathematics and its Applications, Vol. 63, Cambridge Univ. Press, 1996.

[100] WEBSTER, R.: *Convexity*, Oxford University Press Inc., New York, 1994.

[101] WÖLK, R. -D.: Topologische Möbiusebenen, *Math. Z.* **93** (1966), 311-333.

[102] YAGLOM, I. M.: *A Simple Non-Euclidean Geometry and its Physical Basis*, Springer, New York - Heidelberg - Berlin, 1979.

## I want morebooks!

Buy your books fast and straightforward online - at one of world's fastest growing online book stores! Environmentally sound due to Print-on-Demand technologies.

Buy your books online at
**www.morebooks.shop**

Kaufen Sie Ihre Bücher schnell und unkompliziert online – auf einer der am schnellsten wachsenden Buchhandelsplattformen weltweit! Dank Print-On-Demand umwelt- und ressourcenschonend produziert.

Bücher schneller online kaufen
**www.morebooks.shop**

KS OmniScriptum Publishing
Brivibas gatve 197
LV-1039 Riga, Latvia
Telefax:+371 686 204 55

info@omniscriptum.com
www.omniscriptum.com

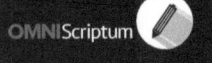

Printed by Books on Demand GmbH, Norderstedt / Germany